MINEWORKERS IN ZAMBIA

For Sue Jones - quite a good trade unionist

MINEWORKERS IN ZAMBIA

Labour and Political Change in Post-Colonial Africa

MILES LARMER

BLOOMSBURY ACADEMIC
LONDON • NEW YORK • OXFORD • NEW DELHI • SYDNEY

BLOOMSBURY ACADEMIC
Bloomsbury Publishing Plc
50 Bedford Square, London, WC1B 3DP, UK
1385 Broadway, New York, NY 10018, USA

BLOOMSBURY, BLOOMSBURY ACADEMIC and the Diana
logo are trademarks of Bloomsbury Publishing Plc

First published in Great Britain by Tauris Academic Studies
An imprint of I.B. Tauris & Co. Ltd. 2007
Paperback edition published by Bloomsbury Academic 2020

A catalogue record for this book is available from the British Library.

A catalog record for this book is available from the Library of Congress.

ISBN: HB: 978-1-8451-1299-8
PB: 978-1-3501-7523-5
ePDF: 978-0-7556-2890-2
ePub: 978-0-7556-2891-9

Series: International Library of African Studies, vol. 18

To find out more about our authors and books visit
www.bloomsbury.com and sign up for our newsletters.

CONTENTS

ACKNOWLEDGEMENTS

This study began life as a PhD supervised by Professor Ian Phimister, who provided greater advice, expertise, and support than I had any right to expect. The first year of research took place at the University of Oxford. St. Cross College provided important support, and the staff of Rhodes House Library was extremely helpful. The Beit Fund generously supported my first research visit to Zambia in August – September 2001. For the remainder of the research, I was based at the University of Sheffield; its Department of History, as well as providing a generous scholarship, supported my second (August 2002 - March 2003) and third (November - December 2003) research visits. Gavin Williams and Megan Vaughan were astute and incisive examiners of the thesis, providing invaluable input that helped shape its conversion for this publication. Adapted sections of the thesis have subsequently been published in the Journal of Southern African Studies and the Journal of Historical Sociology, and I am grateful to those publications for permission to use some of this material in this volume.

I was affiliated to the University of Zambia Graduate Studies Department during my second research visit. I received support and friendship from UNZA's Department of History, particularly Professor Bizack Phiri, Dr Webby Kalikiti, and Dr Yizenge Chondoka. My extensive use of the National Archives of Zambia was greatly assisted by the NAZ staff, ably led by Dr Crispin Hamooya. Drs Marja Hinfelaar and Giacomo Macola of the NAZ's Non-Governmental Archives project, also provided valuable assistance. The archives of the Zambian Consolidated Copper Mines proved an essential source, thanks to the dedication of its staff and Director, Mr Shambweka.

The Zambia Congress of Trade Unions and the Mineworkers' Union of Zambia provided unstinting support and assistance, with the MUZ in particular granting open access to their historical records. In this regard, I am grateful to ZCTU President Leonard Hikuamba and General Secretary Sylvester Tembo, and to MUZ General Secretary Ernest

Mutale. ZCTU Director of Research Austin Muneku, and Charles Muchimba and Dennis Lukwesa of the MUZ's Research Department, provided both expertise and friendship. The officials and staff of MUZ's branch offices were equally helpful in enabling access to records, and assistance in identifying interviewees. My lasting gratitude is offered to the 62 present and former MUZ national and local officials whom I interviewed for this project. They responded to my endless questioning with intelligence and patience. I hope that this thesis does some justice to the communities and workplaces they have represented.

Finally, I would like to thank the following individuals who have commented on various aspects of this research, provided hospitality, or whose support, encouragement and friendship has made the research process both possible and pleasurable: Rita Abrahamsen, Peter Alexander, Chris Bolsman, Carolyn Brown, Sakhela Buhlungu, Nicholas Cheeseman, Penjani Chondoka, Lionel Cliffe, Laura Cole, John Craig, John Darwin, Jo Duffy, Andrew Fairweather-Tall, Ashley Fataar, Patricia Feeney, Keith Flett, Penny Fowler, Alastair Fraser, Maureen Gallaccio, Eugene and Winnie Goss, Anna Gust, Munyaradzi Gwisai, Rick Halpern, Ruth Jolly, Sue Jones, Godfrey Kanyenze, Steve Kibble, Lize Kriel, Vivek Lehal, Hugh Macmillan, Shula Marks, Lindsey Martin, Dunbar Moodie, Horacio Mosquera, Mike Murphy, Kathryn Nwajiaku, Caroline O'Reilly, Deborah Potts, Dave Renton, Nigel Stephenson, Andy Strouthous, John Tanner, John Walker, Eddie Webster, Tracy Wire, Andy Wynne and Leo Zeilig.

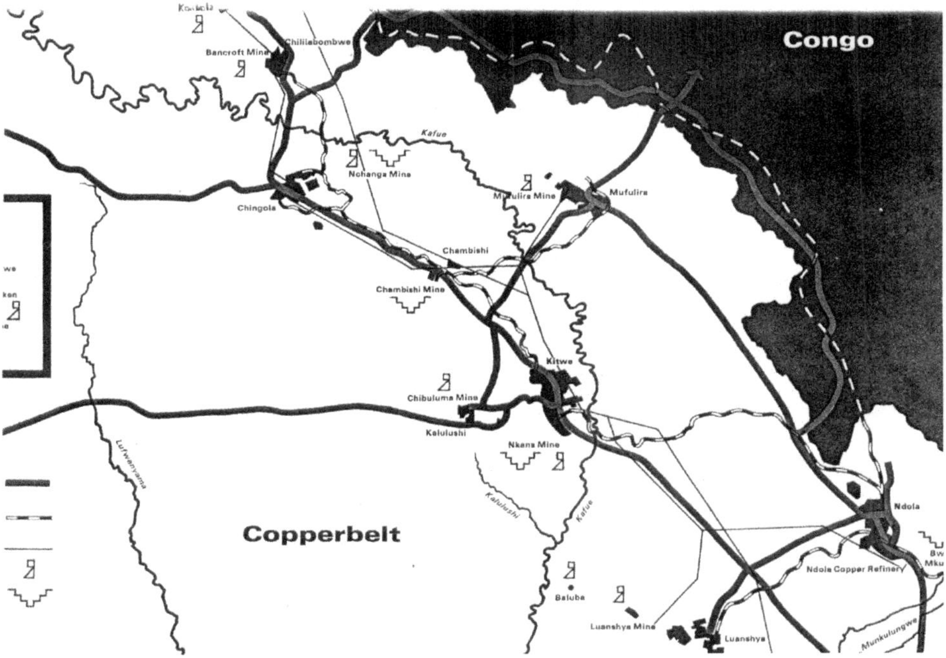

The Zambian Copperbelt

INTRODUCTION

In February 2004, the Mineworkers' Union of Zambia (MUZ) celebrated its fiftieth anniversary at a gala dinner at the Hotel Edinburgh in the Copperbelt town of Kitwe. The golden jubilee's late arrival (the MUZ's precursor, the African Mineworkers' Union (AMWU), was in fact founded in 1950) did nothing to dampen the celebrations, but the event was marked by a mood of wistful nostalgia for a time when the union was a significant power in the land. The guest of honour was former President Kenneth Kaunda, whose presence prompted shouts of joy from mineworkers present, some of whom asked him to come back into power. In his speech,

> Dr. Kaunda said government may not find common ground on all issues affecting workers but they are ordained to work together in the interest of the attainment of industrial harmony and national productivity.[1]

Nostalgia is a powerful force in contemporary Zambia. As James Ferguson has observed, few countries in the world have experienced such a radical decline in livelihoods and living standards. The affection in which Kaunda is now held by mineworkers is, as this study will show, at considerable variance to the highly conflictual relationship between Kaunda's governing United National Independence Party (UNIP), which ruled Zambia from 1964 to 1991, and the mineworkers and their union. It is, however, a reflection of the extraordinary decline of both their industry and their living standards over the last 25 years. This decline began under Kaunda, but rapidly accelerated in the 1990s under the Movement for Multi-Party Democracy (MMD) Government, which successfully ousted UNIP's one-party regime in 1991. The MMD, which came to power with the overwhelming support of the MUZ and its members, presided over a disastrous process of economic liberalisation and privatisation which has led to a halving of the labour force and the closure of some mines. The MUZ has, unsurprisingly, been severely weakened by the industry's decline. Frustration at the union's inability to address job losses, as well as suspicion regarding its complicity in the worst aspects of the privatisation process, has led thousands of mineworkers to defect to rival unions (see Chapter

Six). Once one of Africa's most powerful unions, the MUZ is now close to collapse.

In this context, it is unsurprising that mineworkers who were once in frequent conflict with Kaunda now revere him as an elder statesman. Their attitudes, however, reflect a far deeper confusion, both popular and academic, about the relationship between Zambia's nationalist politicians, the labour movement in general, and the mineworkers' union in particular. Some analysts have portrayed a generally conflictual relationship between progressive nationalism on the one hand, and apolitical, privileged, and reactionary trade unionism on the other.[2] Others have claimed that trade union support for UNIP reflected the fact that unionised workers were part of the party's urban consumer-oriented support base.[3] Virtually all post-colonial analysts have tended to agree that the mineworkers and their union were in some way politically reactionary, or simply apolitical. It is this consensus that this study seeks to challenge.

The prospective reader is entitled to ask what a new study of Zambia's copper mineworkers can tell us that we do not already know. After all, few aspects of African economy and society have been as fully studied as the copper mines of Zambia and their workers. With the start of large-scale mining in the 1930s, sociologists and anthropologists found in the Copperbelt mining towns a new form of urban African society and personality; others found labour migration to the mines to be the primary cause of social dislocation and economic decline in the rural communities that sent men to work there. The centrality of the mines to the national economy attracted the attention of economists who sought ways to improve and increase the exploitation of Zambia's most valuable resources, its copper ore and its people. The apparent overlap between racial and class-based exploitation led many observers to see miners and other workers as the vanguard in a (variously defined) struggle for national liberation and social change. Before formal Independence in 1964, nationalists identified mineworkers as an important weapon in achieving their political aims, and were simultaneously frustrated at their supposed unwillingness to play the role prescribed to them by nationalist ideology.

Mineworkers' alleged failure to fully engage in the nationalist struggle led many political scientists and development economists to see them as a privileged elite. In the years after Independence, mineworkers were accused of undermining state-led economic growth and social development. They were either apolitical or politically reactionary, and therefore represented a brake or block to progressive political change, assumed to be represented by the ruling United National Independence

Party (UNIP). In the 1980s, international financial institutions and their academic allies equally saw the mineworkers and the wider urban population as a consumption-oriented elite, resistant to the necessary liberalisation of the economy.

In striking contrast to this established depiction, the significant role played by the mineworkers and their union in the movement for multi-party democracy that removed UNIP from power in 1991 is widely acknowledged, although until now it has not been directly examined. This study seeks to explain the apparent contradiction between this political participation, and the dominant 'apolitical' perspective described above. It finds that, far from being exceptional, mineworkers' active participation in the pro-democracy movement was in fact the culmination of the consistently influential political actions of Zambia's mineworkers since Independence.

Such an analysis necessarily requires a re-examination of what it means to be politically active in post-colonial Africa. For far too long, 'political' was synonymous with political parties and the problems of governance.[4] This study rejects the narrowing of political ideas and actions to those self-consciously concerned with such institutions. Certainly, some of Zambia's mineworkers were consciously and actively political; they were engaged in organised opposition to UNIP in the early 1970s, and again from the mid-1980s. Others displayed a distinctive political consciousness, in that their participation in industrial and community-based activities designed to improve (or prevent the decline of) their living standards was informed by a widely held sense of their unwarranted exploitation, the inequitable distribution of the revenue generated by their hazardous and harsh labour, and their alternative visions of an independent and developed Zambia. For many mineworkers much of the time, politics amounted to those practical activities that, as Campbell puts it, represent 'the diverse strategies adopted to reduce the insecurities of waged labour…'.[5]

The study finds the basis for this multilayered politicisation in mineworkers' consciousness of the importance of their labour in ensuring the success of the strategic copper mining industry, and through it, national development. Changes in these political ideas, and their practical expression in particular, were affected by the international economic and political climate, the extent of effective political repression, the ownership and organisation of mine workplaces and mine townships, and the structure and leadership of the trade union movement in general and the mineworkers' union in particular. These factors influenced mineworkers' understanding of the forces that have shaped post-colonial political change: the enduring centrality of copper revenue to development and Government policy; the vulnerability of this revenue to international market factors beyond Zambian control; and the growing influence over national political decision-making of the international financial institutions (IFIs), the International Monetary Fund and the World Bank.

Whilst this study identifies a general mineworkers' consciousness, it also rejects the depressingly common tendency to characterise unionised mineworkers as a single homogenous unit that thought and acted uniformly. Mineworkers are, of course, divided by differences in their areas of work, their skills and wages, ethnic affiliation, religion and, more recently, their gender. Most significantly, this study identifies the enduring importance of conflict within the Zambian mineworkers' union. Politically informed divisions between rank-and-file members, local union officials and national union leaders have occurred since the union was founded in the early 1950s. These have arisen from conflicting perspectives regarding the appropriate political role for the union in a post-colonial African context, reflecting the differing interests of the unions' members and leaders.

The social scientists of the Rhodes-Livingstone Institute, who first identified the particular character of Copperbelt society in the 1940s and 1950s (see Chapter One), were generally painstaking in their efforts to understand at first-hand the beliefs and behaviour of mineworkers. In contrast, most post-colonial political scientists and developmental economists have ascribed to mineworkers a role, position, or set of beliefs without having actually talked or listened to them. There are two significant exceptions to this rule: the work of Burawoy, who explored the experience of mineworkers at the point of production; and of Ferguson, who sought to understand the cultural values that underlay their position in urban and rural Zambian society.[6] Both of these studies provide valuable insights into the experiences and perspectives of mineworkers about post-colonial Zambia, and these are analysed below.

Neither Burawoy nor Ferguson sought to systematically assess how the failure of the post-colonial political settlement to meet mineworkers' expectations of social and economic change influenced their subsequent attitudes towards that settlement. This study aims to explain how mineworkers' understanding of their own position in society and economy influenced their actions in ways that had a significant political influence in post-colonial Zambia. It argues that the dichotomy between mineworkers' expectations for liberation, and the reality of their continuing exploitation, is the primary reason why mineworkers continued to demand substantial improvements in wages and conditions, utilising their local workplace and community organisation to do so. In order to maintain this capacity for effective industrial action, they consistently resisted the control of their workplaces, living places, and trade union by UNIP and the post-colonial state that it dominated, in a context of economic decline and an increasing rate of exploitation of their labour.

The study explores the development of this gap between expectations and reality, from the declaration of Independence in 1964 to the return of multi-party democracy in 1991. It explains how, in the years after Independence, mineworkers' political support for UNIP did not prevent them from making demands for the transformative changes in wages and conditions that they expected to result from the end of colonial rule and the establishment of democratic national government. This led them to take industrial action in 1966, bringing them into early confrontation with the mine companies and the state, which detained some of their local leaders and sought to end the independence of the mineworkers' union. In the late 1960s, the creation of a unified mineworkers' union, in the form of MUZ, appeared to deliver the more responsible union leadership sought by UNIP and the mine companies. However, directly elected MUZ branch leaders challenged agreements reached by the union's leadership which they believed failed to meet their members' expectations of an equitable post-colonial settlement. The state's direct intervention in the union's internal affairs to end this dissent clarified the ruling party's position towards their demands. Many mineworkers subsequently supported the political challenge offered to UNIP by the United Progressive Party (UPP) in 1971, before this was suppressed by the declaration of the one-party state.

For the rest of the 1970s the MUZ, and its members, appeared on the surface to have accepted the limitations of the post-colonial political and economic settlement. In practice, mineworkers' demands did not go away. They were rather displaced from official expression via formal industrial relations to more unofficial forms of industrial action, located particularly in the mine townships, which remained outside effective state control. Mineworkers and their wives organised unofficial strikes and protests to express their enduring grievances, unsanctioned by MUZ but utilising its local structures. In 1981, UNIP's attempt to integrate the mine townships into the structures of the one-party state led to confrontation and to the identification of the labour movement (of which mineworkers represented the most significant element) as the effective opposition to UNIP. In 1982, mineworkers replaced an MUZ President they believed was too close to government, and therefore incapable of adequately representing their interests, with a new leader associated with the successful protests of 1981.

In the 1980s, Zambia's economic decline and the increasing influence of the IFIs over government policy led to a significant reduction in mineworkers' real wages and conditions. Mineworkers and their families found that their demands could only be met by (increasingly violent) direct action. In the late 1980s, many local MUZ leaders and ordinary mineworkers engaged in underground political organisation, with the aim of removing UNIP and the one-party state. Despite their inability to prevent declining living conditions, they were able to overcome UNIP attempts to control their union leadership. This culminated in the prominent active participation of both MUZ and its members in the MMD, which succeeded in ousting UNIP in the elections of 1991.

Throughout the period explored by this study, most mineworkers rejected official explanations of the supposed 'realities' of post-colonial governance and development and the ways in which these limited the realisation of their aspirations for themselves, their families and their communities. Instead, they consistently utilised their local workplace and community organisation, and the values on which this rested, to realise these aspirations. In so doing, many of them found themselves in periodic conflict with the UNIP-dominated Zambian state and the international agencies which increasingly defined government policy.

The political importance of mineworkers' activities led UNIP to attempt to ensure their acquiescence by controlling their union leadership. Mineworkers responded by consistently removing leaders perceived to be aligned to UNIP, replacing them with those whose autonomy from the ruling party would, they believed, better enable them to campaign for improvements (and resist reductions) in real wages and conditions. The increasing inability of industrial action to address falling wages, and the state's attempts to control their union led mineworkers to politically oppose UNIP and to support the MMD. In so doing, the majority of mineworkers accepted the surrender of the political autonomy of their union that they had assiduously defended since the 1950s.

Understanding African labour

Modernisation and the 'Urban African'

In the development of African labour historiography in the second half of the twentieth century, the Copperbelt has served as a compelling exemplar of labour's role in political and social change. From the 1930s, the rapid emergence of a region of industry, wage labour and urban settlement in a territory otherwise stereotyped as rural and unchanging attracted the attention of researchers, particularly after the riots and strikes of 1935 and 1940 alerted colonial policy makers and mine companies to the problems of managing a labour force apparently unrestricted by chiefly authority (events during this period are covered in Chapter One). Research by the Rhodes-Livingstone Institute (RLI) achieved significant understanding of the new Copperbelt towns and the social and cultural expression of their African residents.[7] Ferguson has criticised the modernist meta-narrative that he argues led RLI researchers to overstate the extent and inevitability of early urbanisation.[8] It was, however, not primarily inaccuracies in RLI research itself, but rather the political and industrial context in which such findings were received and utilised, that shaped the idea of the urban Copperbelt community. Mine companies resisting wage increases necessitated by urban stabilisation; colonial officials anxious about discontent urban Africans detached from the moorings of 'ancient' rural values; chiefly elites losing control of the labour of youths on which their economic and political power rested: all had an interest in drawing attention to the challenges posed by a new urban African society.

These anxious perspectives established a misplaced framework of understanding that has pervaded studies of urban Africa (significantly influenced by misplaced ideas of the Copperbelt) ever since.[9] The unwarranted bifurcation of rural and urban Africa, with distinct and opposed values, economies and societies, is its most significant legacy. The common tendency to make such a distinction between rural and urban experiences has been informed more by morality than evidence. Political scientists have demonstrated a worrying tendency to accept politically instrumental myths about the enduring and non-exploitative nature of rural African societies and the opposing stereotype of African towns and cities as places of consumption and corruption (see below).

Nationalism and Labour

In the 1950s, diverse challenges to the economic and social impact of colonial rule in Africa generally coalesced into singular nationalist organisations. The political impact of labour movements tended now to be assessed not on their own terms, but on how they related (and contributed) to the particular aims of such organisations. As Freund argues, 'It was assumed that the union was an arm of the nationalist movement'.[10] Freund shows how Comintern influence led many theorists, particularly in the South African Communist Party, to subsume workers' grievances into nationalist struggles, as a first stage of a two-stage revolutionary process.[11] Frederick Cooper's seminal study of African labour in French and British colonial Africa demonstrates how the conflation of all anti-colonial activities into a nationalist meta-narrative undermined the capacity of diverse sections of African society, including workers, to address their particular grievances.[12] Cooper has argued elsewhere:

> In the mid-1950s, a direct clash emerged between the principle of class struggle and African unity [that had] implications … for what kinds of Africas would be imaginable in the postcolonial era and what kinds of Africas would be excluded from political debate … for some, the end of empire meant that impoverished and oppressed people could share in the universal good, could aspire to a generally defined minimum standard of living, could insist on certain rights, as workers, as citizens, as women, as individuals … as African labour movements were among the first to learn, the very insistence on the primacy of 'African' community can constitute a denial of solidarities and differences that people experience in their daily lives.[13]

In practice, researchers such as Berg and Butler found that labour was playing a negligible role in many nationalist movements.[14] In Northern Rhodesia, the colonial Government's aim of keeping nationalist politics and trade union affairs 'in separate compartments' was, according to Mulford, largely successful.[15] It was not only nationalists who were critical of the apparent lack of commitment of labour organisations to their model of progressive political change. Frantz Fanon's critique of emergent bourgeois nationalist leaderships also saw organised labour as part of an emergent African elite in inevitable conflict with the rural poor who, Fanon believed, were the truly progressive class:

> the workers are in fact the most favoured section of the population, and represent the most comfortably off fraction of the people … Any movement starting off to fight for the bettering of living conditions for the dockers and workmen would … run the risk of provoking the hostility of the disinherited rural population … this peasantry precisely constitutes the only spontaneously revolutionary force of the country…[16]

The Limits of National Development

At Independence, such arguments chimed with aspects of new nationalist ideologies, which promoted an idealised view of values that (it was claimed) arose directly from 'traditional' rural African society: a rejection of materialism, the celebration of communal decision-making and labour and a respect for chiefly authority, now re-imagined at a national level in the form of the ruling party and its leader. In Zambia, Kenneth Kaunda's 'Humanism' was an important step in the imposition of such an ideology, seeking as it did to conflate the values of the ruling UNIP party with those of the nation as a whole.[17]

For the Western development advisors who played a significant role in defining post-colonial Government policy, such ideas provided an important ideological justification for what they saw as a necessary dampening of expectations of post-colonial change. Developmentalist approaches emphasised, at least theoretically, the centrality of rural development, and were generally hostile to the supposedly unproductive and uncontrolled urbanisation they believed was inappropriate for post-colonial African development.[18] The contradictions inherent in this idealisation of rural-based development were particularly acute in Zambia. Urbanisation had occurred early, was of an exceptional level by the 1960s and was based on the most productive sector of the Zambian economy, the copper mines.[19] The agricultural sector, in contrast, was one of the least productive in sub-Saharan Africa. Zambian development plans claimed to utilise mine revenue for rural development. In practice, such funds were used inconsistently and accrued primarily to prominent local UNIP leaders, who treated development 'loans' as a reward for their political loyalty. The failure to achieve rural development was laid at the doors of urban Zambians, particularly organised workers, for their unwillingness to sacrifice sufficiently.

Robert Bates' work exemplified such developmental assumptions. His analysis of the relationship between UNIP and the mineworkers' union in the 1960s (which is further examined in Chapter Three) assumed that post-colonial development would be achieved largely through the agency of the UNIP-led state.[20] Any alternative base of authority that undermined the state's capacity to do this would necessarily need to be controlled; the mineworkers' union was a particularly problematic example of this. Bates judged mineworkers' loyalty to UNIP to be largely unconditional, based axiomatically on its role as liberator and government. Mineworkers' loyalty to their union, he suggested, was both more conditional and in decline, now that the state was in a position to address their grievances in regard to racial inequality in the mines. This was based on Bates' personal observations and interviews with politicians, managers, and national union leaders, from which rank-and-file opinion appears to have been largely assumed. This study suggests, in contrast, that most mineworkers' support of both union and party was conditional, based as it was on their respective capacity to articulate and address mineworkers' grievances. In practice, however, it was the union, particularly at a local level, which proved more able to do this.

Zambian workers speak

Bates' perspective was challenged by two studies that provided direct evidence of the attitudes of Zambian workers. Kapferer's detailed analysis of a Kabwe textile factory, carried out in 1966, found that whilst there was significant cooperation between the company's management and their union, workers generally opposed the presence of (outside) union officials during a dispute, assuming this would lead to a negative outcome. They preferred instead to rely on their own local workplace leadership and certainly did not look to UNIP to advance their interests.[21] Burawoy's seminal study of Copperbelt mineworkers provided a strikingly different view of them than that obtained by Bates. Using a participant observation approach that enabled him to study the contested underground production process, he interviewed ordinary mineworkers about their understanding of relations with white and Zambian supervisors and the impact of nationalisation (see Chapter Three). Whereas Bates had suggested mineworkers' primary affiliation was to UNIP, Burawoy's respondents reported their hostility to both UNIP and to ineffective and distant national MUZ leaders. They identified an effective alliance between the Government, the nationalised mine companies, and the MUZ, to increase their exploitation:

> We work harder than we did in the Federal days. Pay scales are low. The Union and the Company work together with Government to bring in pay groups which have cut our categories and acting allowances.[22]

Burawoy's work, whilst highly significant, was limited in two primary ways. Firstly, its emphasis on the production process provided little insight into the dynamics of the mine township community, which this study suggests were at least as important an influence on the political consciousness of Zambian mineworkers as the workplace itself. Secondly, it necessarily provided only a snapshot of labour relations at a time of significant economic and political upheaval. This study, whilst making no effort to replicate Burawoy's detailed contemporaneous analysis of post-colonial industrial relations, attempts to situate the valuable evidence it provides within a wider historical context.

The Labour Aristocracy Debate

The view that urban workers represented a privileged section of society, with interests opposed to those of rural Zambians, persisted into the 1970s. Henderson, for example, claimed in 1972 that 'wage demands by [Zambian] urban workers helped to push up prices, further impoverishing the rural areas', ignoring the fact that prices were primarily affected by structural factors outside mineworkers' control.[23] Pettman provides a particularly characteristic example of such analyses:

> Wage employees in Zambia are not an underprivileged majority, but a privileged minority, and their privileges have increased rapidly since independence, partly as a result of the economic boom in which the rural population has not shared ... African trade union pressure for parity, especially through the MUZ, rapidly reduced the differences between local and expatriate wages, thereby forcing up wages to a point where the economy can no longer support them.[24]

Such a statement assumes that mineworkers' wages were consumed only in urban areas, rather than redistributed amongst their kin, some of whom lived seasonally in the villages. It also assumes the existence of a 'national economy', ignoring both the integration of the Zambian economy into international financial relations and the fact that what could be 'afforded' was a politically contested outcome of a triangular relationship between government, unions and international mining capital.

These views complemented the widespread argument in African labour studies that the continent's organised working-class represented a 'labour aristocracy'. This was put forward most influentially by Arrighi and Saul. Offering a structural analysis influenced by Fanon, they argued that African workers' privileged material circumstances (in comparison to the peasantry) led them to identify with existing forms of economic exploitation.[25] Arrighi and Saul's depiction of Africa's labour aristocracy implicitly reflected their assumption that a revolutionary class existed that they believed would successfully challenge global capitalism. African workers failed to meet their stringent criteria for such a class – those post-colonial industrial stoppages that had taken place were dismissed as 'populist' outbursts that could be easily contained. Saul later conceded there was a danger of equating the working-class too readily with 'elite' groups. He nevertheless maintained that:

> … the most organised and articulate of those proletarianised by the imperial impact appeared to have been 'processed' in such a way as to facilitate their material and cultural identification with the system of neo-colonial domination.[26]

Such comments exemplify the tendency of structural Marxists to see workers as passive agents who were acted upon by political and economic forces entirely beyond their control.

Labour History From Below

The labour aristocracy debate was a dominant theme in post-colonial labour studies in the early 1970s.[27] However, from the mid-1970s, significant shifts in the study of labour in Africa took place. A new generation of South African labour researchers identified the centrality of migrant labour to the economic basis of Apartheid. Exploring the economic interrelationship between sending and recipient areas, they demonstrated the interactive and dynamic nature of rural-urban economic relationships.[28] This influenced a new set of studies that significantly increased understanding of the colonial Copperbelt (the findings of these studies are outlined in Chapter One). Phimister and van Onselen identified hidden and creative forms of resistance by migrant mineworkers to Southern Rhodesian colonial labour controls, indicating the potential for such action amongst workers who had achieved neither urban stabilisation nor formal union representation.[29]

Following Robin Cohen's work on Nigeria, researchers carried out detailed case studies of workers' consciousness and actions in colonial and post-colonial West Africa. Peace's study of Nigerian factory workers in the early 1970s found trade unions in Ikeja to be 'autonomous local level organisations led by democratically elected shop floor men [that were] fully geared to protecting the interests of ordinary workers'.[30] In so doing, they 'presented a severe challenge to those in power'.[31] Crisp's study of Ghanaian gold miners found that they consistently influenced the negotiating position of their union by refusing 'to give anything more than conditional support to the union leadership'.[32] The post-Independence nationalisation of Ghana's gold mines made criticism of management a *de facto* attack on government and increased state control over the mineworkers' union. Such developments, as Burawoy found in Zambia and elsewhere, encouraged a rapid generalisation from economic demands to political ones.[33] Crisp specifically rejected the artificial division of workers' actions into 'political' and 'economic' categories:

> Conventionally, historians and social scientists have tended to draw sharp distinctions between the different arenas of conflict between labour and capital … Workers are said to act 'politically' when they participate in organisations which have a recognized role in the national political process, or when they challenge directly the legitimacy of the state's ruling elite. When they attempt to maintain or increase the level of their wages, they are liable to be described (or even dismissed) as acting in an 'economistic' manner, and when they take action over workplace issues they are said to be acting 'industrially' … the struggle between labour and capital proceeds simultaneously in the workplace, the community, the national political arena, and in the rural labour reservoir. While it recognises that overtly political resistance often has a more profound and more immediate impact on the relationship between labour, capital and state than other forms of action, by redefining 'politics' in terms of the efforts of individuals, groups and institutions to assert and resist authority, this framework rejects the suggestion that 'economistic' and 'industrial' struggles are apolitical in either motivation or effect. Industrial struggles question the organisation and control of a labour process which is dependent on a reflection of a particular configuration of authority relationships, while economistic struggles question the differential distribution of benefits inherent to that configuration.[34]

Jeffries found that Ghanaian railworkers provided leadership to the wider urban poor in protesting perceived exploitation and wider social injustice.[35] Railworkers were suspicious of political ideologies with little immediate relevance to their struggles:

> While the railway workers' culture is highly political … The working assumption of Railway Union political culture is that untrustworthiness and oligarchic self-interest are to be expected of all party politicians and of any union leader who associates too closely with them. There are two things the railway workers will not

> stand for. An untruthful leader, someone who changes his mind to suit his interest. And a leader who sides too strongly with a political party. They know he will sell them out.[36]

In resisting attempts by Nkrumah's Government to control their activities, rank-and-file railworkers utilised a consciously 'apolitical' trade unionism that nevertheless had a significant political impact. When workers protested against tax rises by Busia's Government in 1971,

> … the government characterised the strikers as a relatively privileged labour elite pursuing selfish interests to the detriment of the real poor, the unemployed and the rural populace … Yet it is readily understandable that such an emphasis on the relatively slight differentiation within the ranks of the poor should be interpreted as a calculated attempt to divert attention from the primary and immense gap between the masses and the elite. There was greater validity to the workers' claim to be acting as 'spokesmen of the people' since, with the suppression or buying off of opposition parties, the unions represented the only institutional means of the expression of popular discontent.[37]

In most of these cases, workers were not fully proletarianised, retaining powerful links with rural areas and sharing interests with informal workers in residential proximity to them. In northern Nigeria, Lubeck found that workers utilised religious values as an effective force for industrial mobilisation.[38] He argued:

> As opposed to abstract definitions of structuralist theory, historical interpreters of class analysis are interested in how classes as social and historical actors shape their own history and act out political solutions which determine to a significant degree the internal boundaries and conscious self-identity of their class. Viewed from this perspective classes are situationally determined and are contradictory historical actors, influenced by political and even precapitalist traditions.[39]

The theoretical conclusions of such works were, and remain, significant. Having identified the ways in which the 'labour aristocracy' argument served to strengthen governing elites in their efforts to divide the poor, Jeffries argued that Arrighi and Saul over-emphasised the determining effect of economic position on class consciousness and political action. He warned researchers against searching for a preconceived 'revolutionary' perspective that militated against a more objective analysis of the capacity of workers (and peasants) to effectively challenge their economic exploitation. This was not a rejection of the importance of ideology in African trade unions, simply of its *a priori* narrowing to an abstract notion of revolutionary ideology with little basis in reality.[40] The approaches and arguments of these researchers have significantly influenced the methodology utilised in this study.

The Cultural Turn

This school of colonial and post-colonial labour histories, drawing as it did on E.P. Thompson's 'history from below' approach, gave way in the 1980s to Thompson's other legacy, the micro-study and the rejection of structuralism expressed in post-modern or cultural approaches to African studies. Such approaches have provided important insights into particular African societies, highlighting in particular the previously neglected centrality of gender relations. In Africa, as elsewhere, they presaged a turn away from urban labour. Such studies tended to explore particular cultures and groups without clear reference to the national and international power relations within which they operated. James Ferguson's 'Expectations of Modernity' was, as already stated, a sustained attack on the modernisation narrative that he saw as having inaccurately portrayed the progressive urbanisation of the Copperbelt. Ferguson questioned the accuracy of RLI research, suggesting that the Institute and its successors understated the extent of urbanisation in the 1930s (the supposed period of 'labour migrancy'), and overstated the loss of rural ties since the 1960s. Macmillan's subsequent critique correctly identifies Ferguson's over-generalisation in this respect.[41]

Ferguson is nevertheless correct that, despite high levels of urban migration after Independence, most urban Zambians retained significant social, economic and cultural ties to rural areas. Recent research by Potts supports the view that the size of Zambia's urban population has been overstated, partly by developmentalists alarmed by the persistent myth that more than 50% of Zambians were urbanites. Potts also shows that the Copperbelt's population has been gradually falling since the early 1980s, as economic decline reduced employment opportunities in towns.[42] The Ferguson-Macmillan debate tended to obscure the realities of an urban population that retained significant ties to rural areas. Mineworkers, who were housed by mine companies for the duration of their employment had, until recently, little choice but to retire to their 'home village', despite the fact that they might have spent their entire lives in mine townships. In this sense, depictions (by Bates, Pettman and others) of Zambia as a 'dual economy' that enabled urban Zambians to exploit rural Zambians were misplaced. Rather, there was (and is) a constant circulation between, and interdependence of, urban and rural populations, enabling the exchange of money, people and ideas between towns and villages.[43] Mineworkers hosted rural kin for periods of education and employment in towns. Their wage demands reflected social obligations to these extended families. In these respects, mineworkers' enduring rural links have not undermined their political importance, but have in fact helped define it.

Whilst this study supports Ferguson's conclusions regarding enduring rural-urban linkages, it argues that his emphasis on the cultural 'styles' through which he claims mineworkers understand their relationship to rural areas tends to obscure the economic and social reasons for the maintenance of such ties. It is notable that, with the sale of mineworkers' houses in the 1990s, many mineworkers no longer retire to villages. The popularity of house sales amongst mineworkers suggests that they are, in general, happy to reduce their links to rural areas when the possibility arises (see Chapter Six).

Most importantly, although Ferguson clearly understood the underlying reasons for Zambia's economic decline, he had virtually nothing to say about mineworkers' understanding of the power relations underlying this decline, nor their capacity to organise in order to resist its effects. For example, Ferguson explains that the paucity of pension provision was a primary cause of mineworkers' inability to retire in towns.[44] Although he mentions the Mukuba pension scheme and the 1985 strike that took place over it, Ferguson fails to understand that the strike was one expression of an enduring demand by mineworkers for a pension scheme that would enable a sustainable urban retirement (see Chapter Five).

Such an approach is symptomatic of Ferguson's unduly sweeping generalisations regarding urban Zambians' inability to make sense of their social reality, for example his claim that 'Copperbelt residents in the late 1980s did not inhabit a stable and well known social order. They did not know what was happening to them and did not understand why it was happening. Neither did I.'[45] In contrast, evidence is presented here that shows that, whilst their social order was indeed in a state of flux in this period, many Copperbelt residents had an understanding of what was happening to them and why. This informed the active participation by mineworkers in covert political opposition to the state, opposition that became overt from 1990 onwards.

Economic and Political Liberalisation

In the 1980s, the prominent role played by African labour organisations in resisting structural adjustment programmes led to their renewed characterisation by IFIs as part of an urban elite that benefited from the existing distribution of wealth and power.[46] Applying this argument to Zambia, Bates and Collier argued that UNIP's need to maintain its core urban 'electoral constituency' meant that 'political structures introduced a systematic pattern of bias in economic policy making' and that 'The economic interests that counted were those of consumers in the urban townships. The result was a political bias that defended immediate consumption at the expense of longer-run growth.'[47] This argument, of course, assumed that structural adjustment policies would ultimately lead to such growth – their practical implementation in Zambia under both UNIP and the MMD suggests the opposite. It also failed to explain why UNIP's supposed core constituency played such a prominent role in its removal from power in 1991.

Following these events, political scientists sought to explain the prominent role played by labour in the pro-democracy movement in many African countries (see also Chapter Five). Akwetey, in a comparative analysis of the role of trade unions in the democratisation of Ghana and Zambia, argued that union movements engaged in democratisation initiatives to remove threats to their existence, to defend their autonomy and their right to participate politically, a right theoretically granted by the corporatist one-party state.[48] Akwetey demonstrated that this 'corporatism' was circumscribed in Zambia, because appointments to state-union leadership positions were made on the basis of personal loyalty to UNIP, not on the authority of the leaders concerned. He also dismisses the supposedly consultative structures established in Zambia, correctly arguing they were designed to more effectively control the actions of workers. Akwetey rightly concluded that the greater the effective interaction between union leaders and members, the more defiant the former were towards government adjustment programmes.[49]

Nevertheless, by focussing overly on the formal positions of the Zambia Congress of Trade Unions (ZCTU), rather than the actions of its leaders, Akwetey wrongly claimed that, 'Until late December 1989, the ZCTU's acceptance of the post-colonial institutional-political framework was not in doubt.'[50] In fact, as studies by both Nordlund and Rakner demonstrated, Chiluba's call for multi-partyism in December 1989 was the culmination of 15 years of ZCTU criticism of UNIP policy.[51] This did not consistently seek the removal of UNIP from office (to do so would have invited immediate detention without trial), but presented a challenge which, in a context of increasing economic crisis, exposed the one-party state's incapacity to address Zambia's problems and highlighted the need for its removal.

Rakner stressed the union movement's capacity to maintain organisational autonomy from UNIP as a primary reason why it was able to provide an important political challenge to it. She concluded that, because 'The union leadership has always been under strong pressure from its rank and file branch level membership,' the latter were able to successfully counter attempts by the state to undermine their leaders' autonomy.[52] Nordlund focused on the influence of mass organisation and mobilisation on elite decision-making processes.[53] He described the actions taken by the ZCTU to ensure its autonomy from the UNIP state and argued that Congress offered a particular threat to 'national development', particularly in relation to control of the economically strategic copper mines. He showed the way in which union leaders perceived as too close to UNIP were 'disciplined' by the mass membership, citing the removal of MUZ leaders in 1982 and 1990.

Nordlund concluded that the Zambian union movement's retention of collective bargaining powers ensured its enduring importance to its members and that the 'ZaCTU's [ZCTU's] objective resources - its ability to cause work stoppages effectively halting the flow of export earnings and foreign exchange into the country - were simply too great for the UNIP to challenge.'[54] This is partly correct; these resources, however, rested not with the ZCTU itself, but with the mine labour force.

These studies provide useful insights into the role played by the ZCTU in opposing UNIP and how its organisational autonomy enabled it to do this. However, they do not explore the motivations of workers for their opposition to UNIP and the way in which this influenced the actions they took. Rakner argued that in order to adequately understand the conflict between the state and the labour movement, future studies would need to look at the internal structure of trade union decision-making and the relationship between this and the expression of union members' views at branch level. It is precisely this relationship that this study examines.

Social Labour Studies

The approach taken here has also been influenced by labour studies that have utilised recent historiographical developments in respect of community and gender relations, without losing sight of the political economy within which the events they describe have taken place. Dunbar Moodie's study of South African gold miners explored the impact of labour migrancy on both miners themselves and their sending communities in Pondoland.[55] Whilst accounting for structural factors (such as the compound system) that influenced mineworkers' organisation, Moodie's focus on the dual identities of his subjects as wage labourers and as patriarchal heads of homesteads demonstrated the way in which workers can form an active, unionised working-class whilst retaining close ties to rural kin and land.

Brown's recently published study of the Enugu colliery in Nigeria is one of the most rounded studies of an African working-class yet produced.[56] She provides a detailed analysis of the economic and international political context within which the mine was developed. By placing colliers' own vision of their future at the centre of her analysis, she demonstrates the unpredictable ways in which this informed their consciousness and actions as workers. Enugu workers sought to reduce their economic insecurity by ensuring continuing access to their farms and to shape their labour in ways which ensured this would be possible. Like Northern Rhodesia's copper miners (see Chapter One), they grasped the opportunity provided in the 1940s by the British Colonial Office to introduce 'responsible' trade unions, to develop unions that reflected their own aspirations. Miners' awareness of the economic importance of their industrial labour enabled them to utilise its withdrawal to advance their living standards, action that had a significant political impact.

Klubock's equally detailed study of Chile's El Teniente copper mine provided a meticulous analysis of the interaction between mineworkers' aspirations, a foreign-owned (and later nationalised) mine company, and an apparently progressive Government that nevertheless failed to fulfil mineworkers' expectations.[57] Comparatively high wages did not buy industrial peace, because

> Members of the mining community perceived their high wages and benefits to be both the result of their history of collective struggle and the just reward of decades of deplorable living and working conditions.[58]

In the 1940s, the Chilean Government's dependency on US capital and aid led to the violent suppression of strike action by the state. In the 1960s, the 51% nationalisation of the copper mines was implemented 'with the support of the North American companies and the United States', leaving the existing mine management largely intact (for parallels with the Zambian experience, see Chapters Four and Five).[59] Chilean mineworkers' responses were informed by their own distinctive community-based organisation and their autonomy from their national union leaders.

The work of Klubock and Brown provides a welcome focus on a factor frequently absent from post-colonial African labour studies: the ways in which the international economic order, and the particular form in which peripheral countries are integrated into it, influenced the shape of industrial relations. Mineworkers, in Zambia as elsewhere, followed the fortunes of their industry on the global commodity markets and shaped their demands accordingly. They did not accept that development was limited to the redistribution of relative poverty within national borders. This study seeks to demonstrate the importance of international economic factors in post-colonial political change in Zambia and to explain mineworkers' responses to them.

These works also place gender relations at the centre of the specific class relationships they identify. Klubock and Brown provide important findings regarding the way in which the subordinated position of miners' wives served to reproduce the (male) mineworker in ways that benefited miners and the mine company. This study, influenced in particular by Parpart's identification of the important role of mineworkers' wives in enforcing industrial action in Northern Rhodesia, argues that they continued to be central to major disputes, particularly in the 1980s.[60] This was partly because the illegality of workplace industrial action displaced to some extent the expression of grievances to the mine townships, where state control was weak (see Chapters Four and Five).

The study emphasises the way in which the spatial proximity of workplace and residence was utilised by the mining community to advance its interests. In this context, the wives of mineworkers regularly played a leading role in advancing demands for better wages and conditions, and improvements in social services provided by the mine companies. They also enforced industrial action by ostracizing morally, and sometimes physically, the families of mineworkers who broke strikes. Wives were treated with hostility by the mine companies, which refused to concede to them a legitimate role in their model of industrial relations, and with ambiguity by MUZ leaders, who lacked any formal influence over their actions. Nevertheless, this study is primarily one of the mineworkers' union and its members and does not claim to be a comprehensive social history of the mine townships themselves. A detailed exploration of the gender relationships in mine townships and households would provide a significant additional contribution to our understanding of the importance of social dynamics to the history of the Copperbelt; this is, however, beyond the remit of this particular study.

Methodology and source materials

The historiographical analysis presented above has informed the development of a methodological approach that reflects the strengths of the 1970s West African studies, and the more recent work of Klubock and Brown, to overcome the flaws in existing analyses of Zambia's mineworkers and their union.

For example, the approach taken attempts to overcome a tendency to favour the official voice expressed in the records of the Ministry of Labour, the mine companies, or the mineworkers' union. The public policy declarations of government are examined, but (where possible) so too are private arguments within government that reveal the underlying motivations of politicians and their advisors. The significance of industrial action is not primarily assessed by an analysis of strike statistics, but by the causes of particular strikes and other actions, the importance of which cannot be simply assessed by, for example, the number of work days lost. The statements of union leaders are not assumed to represent the views of their members. The minutes of union meetings are utilised to provide insight into intra-union dynamics, but are not assumed to be an accurate record of what took place. Interviews, whilst not always providing a reliable source of factual information, provide insights into the ideas and attitudes of mineworkers. This approach has informed the use of primary sources, which are described below. Whilst each source has its weaknesses, together they enable the researcher to achieve a comprehensive understanding of the political role of Zambia's mineworkers.[61]

Ministry of Labour records

The records of the Ministry of Labour (MoL), held at the National Archives of Zambia (NAZ) in Lusaka, provide an important insight into the UNIP Government's labour policy, primarily for the period 1964–72. The Ministry was not the sole authority over policy and legislation; indeed, the records demonstrate the increasingly prominent role played by

President Kaunda in labour relations and the relative marginalisation of MoL officials from the late 1960s. The records show that industrial unrest on the Copperbelt in general and its mines in particular was the primary concern of Labour Department officials. Their reports reveal the importance of public meetings as a primary base for the expression of rank-and-file discontent and the normality of conflict between members and leaders of the union. MoL records show the (previously unreported) extent of the conflict between the ZMU and UNIP in 1965-66. They also demonstrate the UNIP Government's constant preoccupation with the 'control' of organised labour. The 1971 Industrial Relations Act, publicly portrayed as a progressive extension of corporatism that created the potential for extensive worker participation in economic management, is shown in these documents to have been shaped by a debate within UNIP regarding the most effective way that legislation could be used to control workers, via their unions.

Sadly, despite an official twenty-year rule governing public access to Government records, almost no MoL files are actually available for the period after 1972. This appears to reflect a more secretive approach by the one-party state. Ministries were given the right to retain records of immediate relevance to their current activities, a clause which appears to have enabled Ministries to retain most or all their important records. It is also the result of a more recent decline in the capacity and willingness of the state to retain its records. Many files transferred to NAZ offices were subsequently destroyed, because they were declared to be of insufficient importance. In the early 1990s, some low-paid NAZ staff stole large quantities of files and sold them as scrap paper. This inevitably limits our ability to understand the one-party state period. Some insights are, however, provided by the NAZ's recent collection of the private papers of retired politicians.

Mine company records

The archives of the Zambia Consolidated Copper Mines (ZCCM) in Ndola bring together ZCCM's own papers with those of the private mining companies, Roan Selection Trust (RST) and Anglo-American Corporation (AAC) and their nationalised successors, Roan Consolidated Mines (RCM) and Nchanga Consolidated Copper Mines (NCCM). Company records provide an insight into ongoing industrial relations with the mineworkers' union. They demonstrate the continuity of these relations from private to nationalised ownership, showing the lack of any significant union or worker participation in management decision-making.

Pay negotiations show the ways in which mineworkers' understanding of international copper prices and national economics influenced union demands, as well as union leaders' perpetual concern regarding the potentially negative reaction of their members to settlements they regarded as inadequate. Other papers indicate the realities of company-state relations: the continuing autonomy of white mine management in the 1970s, despite partial nationalisation; the increasing influence of the World Bank on management decisions in the 1980s; and the use of ZCCM resources to support politically oriented projects in the late 1980s. Whilst the archive is comparatively well organised, the privatisation of ZCCM has raised doubts about its long-term future. The minority state mine shareholder, ZCCM – Investment Holdings, now owns the archive, but lacks funds to guarantee its long-term future.

MUZ records

The Kitwe-based head office of the MUZ, Katilungu House, has no organised archive of its records, few of which survive from the colonial period. The relatively poor state of MUZ records reflects the recent decline of the union, following mine privatisation and the consequent reduction of the mine labour force (see Chapter Six). Complete sets of the minutes of the MUZ's governing bodies were compiled from the records held at branch offices across the Copperbelt. These are highly uneven; Luanshya's records are particularly impressive, dating back to the 1950s, whilst most of Mufulira's records were destroyed during the 1985 Mukuba pension strike, when the branch office was burned by striking miners (see Chapter Five). MUZ records illustrate the policy positions adopted by the union, towards both mine companies and the state. They also provide an insight into the internal dynamics of the union, demonstrating particularly the perpetual conflict between national and branch leaders. Some branch records include minutes of Branch Executive Committee meetings, at which shop stewards and branch officials expressed their distinctive views of union policy. There are also limited reports of public meetings.

Newspapers

A further significant source of information for this study was provided by Zambian newspapers, particularly the *Times of Zambia* and its predecessor the *Northern News*. Newspapers provided a detailed record of significant events during the period under examination, particularly regarding the national political context, and relations between the state and the union movement. The *Times of Zambia* frequently adopted the official position of the Zambian state towards the labour movement; this increased following the declaration of the one-party state, particularly after its effective takeover by the state in 1982. Its editorial position is therefore periodically used to illustrate the state's public position. Despite this, its news coverage generally provides a useful record of specific events and is utilised accordingly.

Interviews

From October 2002 to March 2003, I carried out 62 extensive interviews (averaging 90 minutes in length) with former MUZ national officials, branch officials, and shop stewards. Although efforts were made to ensure that this group included all age groups, working and union experiences and reflected the general balance of men and women in the mine labour force, no claim is made that they are a representative sample, either of union officials or of mineworkers as a whole (biographies of interviewees are provided in Appendix One). Many former officials are deceased, or returned to rural areas on retirement, making it impossible to secure a sample that would meet sociological standards. For these reasons, few interviewees were available with experience of the 1960s; interview material is consequently used more extensively in chapters covering later periods of the study. Interviewees are treated as historical actors rather than sociological subjects, whose actions have helped shape the events described and whose opinions are cited in so far as they aid our understanding of these events. Individual perspectives are not treated as representative of mineworkers as a whole, but quotations have been used when they are believed to be illustrative of wider opinion amongst mineworkers active in the union.

These interviews provide important insights into the cultural expression of labour discontent and resistance. This includes the central importance of the outdoor union 'public meeting', attended by mineworkers, their families and other mine township residents.[62] These meetings decisively shaped local and national organisation. Their importance illustrates the extent to which MUZ, the power of which undoubtedly rested on the economically strategic copper mining industry, should nevertheless be understood as much as a social or community organisation as an industrial one. Mineworkers' wives are virtually invisible in official papers, but their importance in industrial action was consistently emphasised by interviewees.

Structure of the study

The arguments advanced in this study are put forward in six main chapters. This Chapter introduces the main themes of the study, outlines the historiographical basis of the particular role of Zambia's mineworkers and the wider study of labour in Africa. It also explains the methodological approach adopted in this study and the consequent use of primary source materials.

Chapter One describes the role of mineworkers in political change in colonial Northern Rhodesia. It describes the emergence and economic importance of the Copperbelt and its mineworkers. It argues that, despite official attempts to ensure that mineworkers' industrial organisation was moderate and non-political, it proved in practice to be neither. Although AMWU was politically influential in the 1950s and early 1960s, it was unwilling to follow the lead of nationalist parties, particularly the African National Congress (ANC) and later UNIP. This reflected the different political positions of mineworkers and their union leaders respectively; and the diverse expectations of mineworkers and nationalist leaders for national independence. This, it is argued, created a legacy of distrust for the union amongst UNIP leaders, who unsuccessfully sought to control the organisation of the mineworkers' union before and after Independence.

Chapter Two provides an outline of the post-colonial political and economic development of Zambia. It explores the economic context within which Independence was achieved, analysing the Government's failed attempts to reduce dependency on copper. The complex interaction between regional context and national politics is addressed, in particular the ways in which the undoubted threat posed by Zambia's hostile neighbours was used by Kaunda to justify the suppression of internal dissent. It explains the way in which developmental economists, mine capital and state ideology combined to de-legitimise the post-colonial aspirations of mineworkers and wider urban labour. The failure of the post-colonial state to meet the aspirations of most Zambians led to provinicially-based political conflict within and outside UNIP, culminating in the declaration of the one-party state in 1972.

In retrospect, this was a less important event in Zambian history than the global economic crisis that shortly followed. The rise in oil prices, and the collapse in the cost of copper, destroyed the assumptions on which Zambian development had been structured. Whilst the Zambian economy has been in consistent decline ever since, the question of who would pay for this crisis became ever more politicised. In the late 1970s and early

1980s, Zambia became both poorer and more unequal. At the same time, Zambia's rising indebtedness increased the influence of the IFIs over the economy. They, like their developmental predecessors, advocated a reduction in copper dependency, yet saw copper mining as the vehicle for diversification. They also sought to persuade organised workers and the wider urban population to accept lower living standards.

UNIP's control over Zambia was consistently weakened by the conflicting pressures of IFIs on the one hand and the increasingly active resistance of urban Zambians to structural adjustment policies on the other. Food riots and strike action in 1986-87 successfully derailed one structural adjustment programme. By 1990, food riots and industrial action combined with calls for multi-party democracy. As demands for political and economic change became intertwined, the latter were subsumed in the popular desire to oust Kaunda and bring the MMD to power. The contradictions inherent in the MMD's coalition of business, union, and political interests were to work themselves out only after the 1991 elections.

Chapters Three to Five present the major findings resulting from primary research carried out for this study. They cover the period under study (1964-1991) in chronological order, divided into distinct economic and political periods. Each chapter follows a similar structure, which enables an understanding of the relationship between the major factors influencing the role of the mineworkers in political change. These chapters provide an analysis of the relationship between the UNIP-dominated state and the national trade union movement, primarily represented by the ZCTU. These sections demonstrate the state's attempts to control the trade union movement by shaping the union congress as a powerful central body with extensive controls over member unions. This strategy was initially successful, but the election of Congress leaders who sought to maintain the autonomy of the labour movement paved the way for the ZCTU's increasingly critical position towards UNIP and its policies in the late 1970s and 1980s. The emergence of the ZCTU as the effective opposition to UNIP is described, and the relative failure of UNIP's strategy is demonstrated, as is the way in which the ZCTU provided significant organisational support and leadership to the pro-democracy movement which became the MMD.

Each of these chapters examines in detail the particular role of the MUZ (and its predecessors) and its membership, in relation to the forces detailed above. The initial desire of the union leadership to maintain its autonomy from UNIP, the state, and the trade union congress is examined, as are attempts by UNIP to control and/or replace the mineworkers' union. The establishment of the MUZ is described and the attempts of its leadership to establish a rational, technical approach to industrial relations are explained. The post-colonial expectations of mineworkers are surveyed, as well as the ways in which these led to industrial action in the years after Independence. These expectations led to conflict within the MUZ in 1970-71, which was only resolved with the explicit intervention of the state on behalf of the union's leadership. This appeared to have delivered practical control of the mineworkers' union to the one-party state.

Such an analysis is shown to neglect the importance of mineworkers' local community and workplace organisation, their capacity to resist political control and to organise independently of the national MUZ leadership. These factors played a central role in the protests of 1980-81 against the integration of the mine townships into local government. These protests provide evidence of the importance of local organisation and the widespread disillusionment of mineworkers with the outcomes of national independence. The increasing violence of the strikes and protests of the mid 1980s demonstrated the inability of the state to contain mineworkers' expectations and led to increased participation in underground political activity. Mineworkers successfully overturned UNIP's co-option of their union leadership and were in the forefront of organising the MMD on the Copperbelt. They believed that political change and democratisation would enable the realisation of their enduring expectations for improved wages and conditions and equal pay with their white counterparts.

Chapter Six studies developments in Zambia since 1991. It argues that the central role of the labour movement in the pro-democracy campaign of 1990-91 led to the effective surrender of the MUZ's political autonomy to the new MMD, undermining its capacity to play a watchdog role when the MMD, in a context of severe economic dependency on the IFIs, implemented policies against the interests of organised labour. It surveys the impact of ZCCM privatisation, arguing that mineworkers' interests were neglected in favour of the requirements of international mine capital and the IFIs. It records interviewees' reactions to these events, finding that mineworkers are now divided as to whether they and their union should have provided political support to the MMD. It also examines their views regarding the general relationship between labour and political change, finding a widespread (but by no means universal) syndicalist perspective, with many mineworkers believing that their union can best represent their interests by refusing direct support to a particular political party and instead combining a watchdog role with a focus on industrial grievances.

The Conclusion provides a summary of the principal arguments advanced in this study. It also suggests its wider relevance in aiding understanding of the role of labour in influencing political change in post-colonial societies.

1

MINEWORKERS AND POLITICAL CHANGE IN NORTHERN RHODESIA, 1935 - 1964

Introduction

This chapter examines the role of mineworkers in political change in Northern Rhodesia, from the start of large-scale copper mining in the 1930s until Independence in 1964. It briefly describes the establishment of the industry and its workers' early industrial organisation and action. Such activity was initially suppressed, but in the 1940s colonial authorities sought to establish 'responsible' and non-political African trade unions. African mineworkers nevertheless utilised unionisation to achieve substantial improvements in their pay and conditions, frequently using strikes to do so.

The political role of the union movement was more ambiguous. Whilst colonial and company officials identified a dangerously politicised labour movement, dominated by the African Mineworkers' Union, nationalist politicians found AMWU unwilling to follow their lead in organising industrial action for political purposes. This significantly influenced the relationship between the mineworkers' union and nationalist organisations. Henry Meebelo, author of the most detailed history of colonial labour relations, articulated UNIP's position:

> the AM[W]U, the wealthiest, the best organised and the most powerful African trade union in the country was, for all its might and its strategic position … apparently too inward-looking to play the rightful political role against colonial exploitation and oppression.[1]

The belief that Zambia's mineworkers and their union failed to fulfil their 'rightful political role' (i.e. that prescribed for it by nationalists) led UNIP to distrust AMWU and to seek effective control over it. As this study will show, this nationalist perspective decisively influenced the subsequent characterisation of Zambia's mineworkers as generally apolitical and economistic. In contrast, this chapter suggests that unionised workers and mineworkers in particular had their own distinct experience of colonial exploitation and oppression, overlapping but not identical with that of nationalist politicians.[2] Mineworkers, whose political position was not the same as that of the AMWU leadership, generally supported the most radical positions and activities of nationalists, but resisted the subordination of their union to nationalist organisation and priorities.

The birth of the Copperbelt

In the 1930s, the colonial backwater of Northern Rhodesia was transformed into a major global producer of copper, a vital component in emergent manufacturing and electronic industries. Multinational mining companies, primarily the US-based RST and South Africa's AAC, rapidly expanded production in the remote 'Copperbelt' region. Five large mines were producing 116,634 tonnes of copper by 1936, rising dramatically to 268,551 tonnes in 1941.[3] The recruitment of sufficient and appropriate labour supplies was an immediate concern. Skilled white labour was attracted from South Africa and elsewhere only by relatively high wages. Meanwhile, tens of thousands of unskilled African miners were recruited, drawn into the cash economy through a combination of taxation and a growing desire for consumer goods. From 7,459 African mineworkers in 1933, the black workforce doubled to 14,023 in 1936, reaching 26,023 by 1940.[4]

In the 1930s, global depression led to falling copper prices and reductions in the wages of both whites and Africans. As demand for copper revived, particularly in the years before World War Two, white miners were able to secure a *de facto* colour bar separating 'White' and 'African' jobs. A system and ideology of migrant labour was already well established in Southern Africa: Africans were objectified as essentially rural and tribal, working only for 'target' wages. They were not allowed to settle permanently in the new mining towns, where they might become 'detribalised' and develop new identities which would undermine colonial control. The migrant labour system passed the costs of labour reproduction onto rural African societies and served to justify low 'single male' wages, which helped make Northern Rhodesian copper production amongst the most profitable in the world.

However, the weak colonial state was never able to fully enforce migration controls. Many Africans established themselves as effective permanent residents of the mine towns, though their wages were insufficient for urban reproduction. The mines' requirement for increasingly skilled mineworkers also promoted effective African stabilisation, without any official acceptance of permanent African residence in urban areas. Such contradictions were of secondary importance to the British colonial administration. What mattered most was the continuing flow of strategically important Northern Rhodesian copper, which accounted for 13.42% of non-Communist global production in 1938.[5] A 1937 British Cabinet Paper declared:

> These mines may before long be the governing factor in the copper markets of the world, and their output will become essential to national defence in the event of war and of primary importance to the empire's trade and manufacturing in peace time.[6]

Strikes and stabilisation

Colonial neglect of the African labour question was decisively exposed by the mine strikes of 1935 and 1940. An increase in poll tax sparked the 1935 strike; organised through the Watch Tower sect and ethnically-based Bemba dance societies, it was ultimately put down by force, with six miners being killed. Although attempts were made to portray the event as an ethnically-based riot, the subsequent Commission of Inquiry heard from a Compound Manager that 'These people form a definite social group quite shorn of anything tribal, they live in a world entirely different to the other natives'.[7] Parpart argued:

> The daily experiences at work and in the compounds led to a rapid understanding among the miners of both their common interests and their exploitation by management; despite plans to return them to the rural areas, the stabilisation policy of the mines created a group of workers who increasingly understood that collective labour action was their most effective weapon against capital.[8]

Whilst the protest was channelled through Bemba cultural movements, it was essentially a collective response to falling living standards resulting from the poll tax increase, itself reflective of international economic problems arising from the Depression.[9] Following the strike, colonial officials attempted, ineffectively, to stem the practical stabilisation of the mine labour force.[10] Africans increasingly evaded relatively inefficient colonial pass laws, migrating between mines to ensure continued urban employment. Commercial realities defined the companies' practice: the unattractiveness of Copperbelt employment, relative to the better-paid

South African mines, led them to tacitly accept the settling of African miners' families on the Copperbelt in the late 1930s.[11]

Colonial anxiety primarily focused on white mineworkers, who brought to the Copperbelt the South African practice of militant, but racially-based, labour organisation. They established the whites-only Northern Rhodesian Mineworkers' Union (NRMWU) in 1936. In 1940, NRMWU members carried out effective strike action, which won a substantial wage increase. This provided an instructive example to African workers, who struck immediately afterwards. African miners demanded ten shillings a day, that they should be given jobs reserved for whites and that they should be paid the same wage as whites. 'Tribal' representatives recognised by the mine companies were unable to control the dispute, which was apparently led by an elected 'Committee of 17'. Henderson, however, provided evidence that this was a front for the real leaders who did not wish to be identified. The strike was well organised, but a confrontation between 3,000 strikers and 150 working men was put down by troops; 17 were killed and 60 injured. This episode placed African miners at the centre of official concern. A mine compound manager told the subsequent Inquiry:

> The skilled native is the man who stands on the threshold of the enjoyment of the luxuries of our civilisation, he is the man who is most affected by any rise in the cost of living … he is also the person that we can expect industrial trouble from in the future.[12]

Colonial anxiety regarding the emergence of such 'detribalised' urban Africans prompted major sociological and anthropological studies by the new RLI. Its researchers identified a new urban African society on the Copperbelt, experiencing and responding to radical social upheaval.[13] Epstein suggested that mineworkers had adopted urban and modern perspectives and behaviour.[14] Having lost their respect for chiefly authority, such workers therefore needed representation by modern organisations such as trade unions. RLI research helped persuade colonial officials that new forms of urban authority would be needed to manage African affairs.

Labour organisation

The mine companies, however, initially resisted attempts to establish African trade unions. Tribal Representatives, first appointed in the early 1930s, were established in all mines in 1941-2, as an alternative to unions.

The companies and the colonial authorities, however, feared above all the creation of a unified trade union representing both African and European labour. Whilst the racism of most white mineworkers made this unlikely, African 'junior' branches of the NRMWU were established in 1946.[15] The major Rhodesia-wide railway strike of 1945 provided an additional warning of the rising tide of unorganised African labour militancy.[16] These events appear to have shaped a newly 'enlightened' company attitude, accepting the need for African unions.

The post-war British Labour government, influenced by new theories of industrial relations, sought to address and incorporate increasing African demands.[17] In 1947, William Comrie of the British Trade Union Congress was sent to Northern Rhodesia to help establish African unions along what were misleadingly described as 'British lines', i.e. as non-political bodies which would negotiate over wages and conditions, generally without recourse to industrial action. AMWU was established in 1949 and a Trade Union Congress (TUC) in 1951. Most AMWU officials were initially clerks and supervisory 'boss boys', whose interests were not necessarily the same as the majority of unskilled miners. However, the reservation of 'white' jobs frustrated the aspirations of skilled Africans, who therefore sought to advance their interests via a (problematic) alliance with the mass of ordinary miners. As Parpart notes, 'The numerical preponderance of less-skilled miners kept the union leaders from adopting an exclusivist orientation and all but forced them to stress solidarity regardless of ethnic or occupational identities.'[18]

The Government's role in establishing trade unions led some researchers to see them primarily as an imposition from above, limiting their capacity to represent African workers' aspirations. Henderson argued that 'It is practically impossible to find any evidence that the mass of the African workers wanted trade unions until such time as they came into existence.'[19] However, whereas Africans rejected labour organisation based on ethnicity (see below), they embraced the new unions enthusiastically. AMWU membership rose rapidly, from 50% of the 37,000-strong labour force in 1949, to 80% in 1953.[20]

The establishment of African unions, however, did not achieve the improvement in industrial relations sought by the state and mine companies. Between 1949 and 1953, 30 disputes and 12 strikes occurred, as mineworkers took the opportunity presented by unionisation to express grievances, in a context of sustained high copper prices.[21] AMWU's effectiveness was demonstrated by a three-week long strike to support a wage demand in 1952. 79% of mineworkers voted in favour of the strike, which was peaceful and disciplined. The subsequent Guillebaud Commission awarded significant wage increases.

AMWU members voted overwhelmingly to abolish Tribal Representatives in 1953.[22] Henderson argued that, 'In the event … African Trade Unions became independent bodies used by Africans themselves for collective bargaining, which … involved demands which were closely aligned to those of the African political movement.'[23] As was to be the case with the ZCTU after Independence (see Chapter Three), trade unionists took bodies imagined by their creators as mechanisms of control and remade them in their own interests.

AMWU culture was dominated by an emphasis on branch and shop steward organisation and democratic accountability. Epstein described monthly branch meetings attended by 8,000 miners where 'rank and file members are free to participate in the discussions'.[24] Powdermaker's contemporaneous research in 1962, and Parpart's later work, demonstrated the importance of community organisation in the mine townships to effective industrial action.[25] Parpart showed that women were crucial in mobilising support in strikes, for example preventing 'scabbing' by unemployed men: 'Union leaders deliberately mobilized women both during and between strikes, knowing full well that their support was essential for successful collective action.'[26] As subsequent chapters will show, the role of women remained vital to the mining union's practice in post-colonial Zambia.

Unions and nationalist politics, 1948-61

African unionisation preceded nationalist organisation in Northern Rhodesia, which commenced in 1948 with the establishment of the Northern Rhodesian Congress (later the ANC). There was initially a close relationship between nationalists and trade unionists; a 1951 Labour Department report cautioned:

> The leaders of the African trade unions were all to a greater or lesser extent concerned with politics and there were periods when political considerations kept union leaders from devoting a sufficient time to normal trade union functions.[27]

The primary impetus to nationalist organisation was the Central African Federation (CAF), established in 1953. Federation was predicated on the redirection of copper mining revenues to the benefit of white settlers, mainly in Southern Rhodesia. The ANC protested against Federation, but

AMWU President Lawrence Katilungu refused to support these protests with industrial action. This event decisively influenced subsequent nationalist-union relations. Bates claims that, 'Throughout the nationalist period, the union [AMWU] withheld a major resource from the nationalist parties; it refused to employ the strike weapon in support of political objectives.'[28] This argument reflected the nationalist discourse, which retrospectively reduced all 'political objectives' to those of organised African nationalism. As Henderson argues, 'there was a genuine miners' consciousness which did not necessarily rule out a wider political consciousness, but which was keenly aware that the miners owed nothing to the political elite.'[29] Furthermore, the attitudes and behaviour of the emergent union leadership, encouraged by Comrie to adopt a moderate apolitical unionism, were not the same as those of the rank-and-file membership. Whilst Katilungu refused to sanction industrial action in 1953, hundreds of mineworkers, particularly in Mufulira, did strike and were dismissed as a result.[30]

In the early 1950s, when AMWU industrial militancy was at its height, prominent local branch leaders such as Robinson Puta and Jameson Chapoloko were close to the ANC leadership. They mobilised a strike by Nkana mineworkers to protest the deportation of white nationalist leader Simon Zukas in 1953 (an action opposed by Katilungu).[31] Mineworkers voted in 1954's AMWU branch elections for militant and politically-minded leaders like Puta and Sylvester Nkoma. In 1955, these leaders pressured Katilungu to call a strike with political and economic aims: the pay rate of 10s 8d demanded would have qualified many African mineworkers for the income-related electoral franchise for the first time.[32] 30,000 of 34,000 AMWU members voted for action. The resultant strike lasted two months, and was marked by organised picketing and alternative food supply networks organised by the union.[33]

Katilungu was not apolitical but politically moderate, and this moderation extended to industrial matters (see below). Most union officials were more radical. In 1956, of 59 branch officials, 32 were members and 20 were officials of the ANC.[34] Five of the nine members of the ANC's Supreme Action Council were trade unionists. AMWU leaders also participated in local government structures – in 1954, four of the 16 members of Luanshya's Urban Advisory Council were local AMWU Branch Executive members.[35] Henderson argued that this reflected an effective alliance between union activists and nationalists:

> So long as Europeans were in power, politically and industrially, the alliance survived. There was a sense in which all protests were protests against the entire white power structure. All Africans had in common poor wages and conditions, and a social and economic colour bar.[36]

The politicisation of the wider union movement was evident in the Associated Chamber of Commerce's warning in 1954 that 'African trade unionism is drifting rapidly towards political and nationalistic domination'.[37] The following year, the TUC established a subcommittee to campaign on political matters with open links to the ANC.[38]

In response to rising industrial militancy, the mine companies encouraged the establishment of a separate white-collar African union, the Mines' African Staff Association (MASA), in 1954. This aimed to split skilled union leaders from the unskilled membership and to co-opt an African elite as a counterweight to rising expectations of economic and political change amongst the mass of general mineworkers. Concessions to this group were initially possible because of the ongoing expansion and profitability of the industry. However, despite attempts to make MASA membership compulsory for salaried staff, many remained in AMWU. MASA members were isolated and publicly abused by miners' wives.[39] Most senior miners understood the limitations and political consequences of the concessions offered by the mine companies. One prominent union leader explained, 'Even the more educated miners ... knew the workers must join together to fight, otherwise they would not succeed'.[40] MASA never recruited more than 10% of its potential membership.[41] Its unintended impact was to reinforce the principle of unity amongst mineworkers.

In 1956, in protest at attempts to make MASA membership compulsory for relevant grades, AMWU organised a series of strikes across the seven mines. £3 million of revenue was lost before a Copperbelt-wide State of Emergency was declared. Fifty-five miners, 45 of them AMWU and/or ANC officials, were arrested and 'deported' to rural areas. Katilungu was however not detained and sought to persuade strikers to return to work.[42] With the detention of radical branch leaders, Katilungu re-established his dominance of AMWU. The Labour Department's 1956 Annual Report reported that, 'Since the detentions Mr L.C. Katilungu, the General President, has begun, with Government advice, to build up a healthier and more responsible organisation.'[43]

In order to re-establish the payment of union subscriptions by the mine companies, Katilungu agreed in 1958 to the amendment of AMWU's Recognition Agreement, preventing the use of union funds or organisation for political purposes. In 1959, AMWU agreed that it would restrain its officials or members from supporting any illegal society (including nationalist parties).[44] Parpart correctly identifies a decline in direct AMWU participation in nationalist politics from the mid-1950s, but overstates the extent to which this reflected the views of ordinary mineworkers. A dramatic fall in the copper price from 1956 undermined the union's previously strong negotiating position, but also limited resources available to co-opt an African 'elite'. Many AMWU branch leaders and rank-and-file members continued to be active in nationalist organisations. There is no evidence that the relative de-politicisation of AMWU was popular amongst mineworkers; indeed, union membership fell during this period.[45] Economic discontent also spurred the radicalisation of the wider union movement. In 1957, the TUC's 'Draft Programme' called for self-government, self-determination and a national programme of socialism. It specifically demanded free education, crèches, the nationalisation of mines and transport services and full employment.[46] This position was however moderated when Katilungu became TUC President in 1957.

Unlike the ANC, the breakaway Zambia African National Congress (ZANC) and its successor, UNIP, established a successful grassroots organisation, with many of its strongest sections on the Copperbelt.[47] Mineworkers joined UNIP in their thousands. Younger AMWU branch leaders criticised Katilungu (now an ANC member) and were particularly active in UNIP. As Jacobs pointed out:

> Katilungu's presence in the ANC did not mean rank and file [AMWU] support for that party. In fact, the radical and sometimes violent activities of the ZANC seem to have drawn the support of most workers, and certainly most of the Copperbelt workers.[48]

In 1960, Katilungu's acceptance of a position on the Monckton Commission, established to review the CAF (but boycotted by UNIP), led to attempts to oust him as TUC President. These failed and instead Katilungu expelled most other unions from the TUC; the latter went on to form an alternative Congress. Katilungu was however removed as AMWU President that December and replaced by John Chisata. Meebelo subsequently claimed that 'AM[W]U, under the leadership of John Chisata and Matthew Nkoloma respectively, showed no immediate signs of softening its attitude to UNIP.'[49] This is misleading: Chisata, unlike Katilungu, was a UNIP supporter with a record of supporting nationalist activities. He was, however, not prepared to sublimate AMWU to UNIP dominance.

AMWU now rejoined a unified United Trade Union Congress (UTUC). Unity appeared to have been achieved, and with it the loyalty that UNIP sought from the labour movement. The UTUC formally affiliated to UNIP in 1961.[50] Its' paper *Workers' Voice* proclaimed:

> To the majority of workers who make up the UTUC only one political party has clearly and explicitly stood for immediate self-government and independence, and this is UNIP ... The UNIP derives much of its strength from workers and peasants ... its policy of self-government now represents the voice of the people, 90% of whom are workers. The interests of UNIP and UTUC are complementary. The survival of one is the survival of the other.[51]

In 1962, Matthew Mwendapole, UTUC General Secretary, spelt out why trade unions needed to support UNIP:

> There can be no freedom of Trade Unions in a country still under foreign control. Trade Unions are free democratic institutions and can only thrive and perform their legitimate functions under conditions born out of freedom and democracy.[52]

Mwendapole insisted, however, that trade unions would continue to be needed to safeguard the interests of workers in an independent Zambia.

UNIP and unions, 1962 - 1964

The problematic relationship between unions and nationalists nevertheless persisted. Meebelo described UNIP efforts to achieve trade union unity, but in practice it was only interested in unity on its own terms. Young UNIP supporters with little experience in industrial matters were dispatched to the Copperbelt to form new unions loyal to the party, often by splitting existing ones. Many unions were vulnerable to such manipulation because they were poorly organised and resourced, dependent on funding from abroad and lacking in effective internal democracy.[53] UNIP plans for a general strike, to be unleashed if force was needed to bring about Independence, were rejected by AMWU in January

1962. Revitalised under Chisata's leadership, AMWU organised pay-related strike action that May.[54] UNIP pressurised mineworkers to call off the strike so as not to destabilise elections; they did so, amidst much bitterness.

As self-government approached, AMWU protested its exclusion from UNIP's talks with AAC and RST regarding the future of the mining industry.[55] Such occurrences reinforced its desire to maintain autonomy. Chisata declared that, 'We want the union to remain independent whatever the Government, black or white … I would rather resign than go into a political trade union.'[56]

Proposals for a new UTUC Constitution, creating a powerful and centralised Congress, led in December 1962 to a new split when AMWU and other unions left to form the Zambia Trade Union Congress (ZTUC).[57] Without AMWU funds, the UTUC remained dependent on funds from the International Confederation of Free Trade Unions (ICFTU) and the maladministration that accompanied such dependence.[58] Rival ICFTU and World Federation of Trade Unions (WFTU) affiliations created new divisions in the UTUC in 1963-4. Such problems reinforced AMWU's initial determination to remain outside the UTUC, arguing that many of its affiliated unions existed only on paper.[59] Unification was only achieved in June 1964 (see Chapter Three).

UNIP remained unhappy with AMWU's autonomy. As Meebelo put it, 'Because of its lukewarm support for the independence struggle and its arrogantly and stubbornly anti-UNIP political stance ... the African Mineworkers' Union had incurred the wrathful indignation of UNIP and the UTUC…'[60] UNIP's failure to control AMWU led to an attempt to create a rival union under its control, via the manipulation of MASA (see Chapter Three). Mineworkers, meanwhile, increasingly understood that they could not simply rely on political independence to meet their expectations. In the words of one branch official:

> … those in politics have their responsibility to the nation, but we in the trade union movement have our responsibility to the cause of the workers. Surrendering our responsibility because of political misdirection means sacrificing that cause to the hungry capitalists.[61]

UNIP was still expressing concern about the continuing militancy of the mineworkers on the eve of Independence:

> Unless immediate remedial steps are taken to curb the present trend of things in the labour movement, the Copperbelt can be expected to be a hot bed of trouble as far as strikes occurring with disturbing frequency.[62]

Both AMWU and UNIP expressed mineworkers' central demand for Independence, the advancement of African miners into jobs previously reserved for whites. The 1962 pay agreement established a single pay scale for European and African mineworkers, but a racially-based dual wage structure was reintroduced in early 1964. Whilst mineworkers' wages had risen steadily since the establishment of the AMWU, the absolute gap between these and European workers' pay was greater than ever.[63] The 1962 Morrison Commission identified two distinct definitions of advancement. The mine companies envisaged individual Africans being promoted to higher grades with higher pay; the AMWU wanted significant increases for all African workers to narrow the racial wage gap. Differing perspectives on what constituted African advancement would be a key theme of industrial discourse and action in the years to come.

Conclusion

In colonial Northern Rhodesia, mine companies and colonial officials sought to achieve effective control over the labour power of African mineworkers and the profits they generated. African mineworkers successfully undermined colonial labour controls and used industrial action and organisation to achieve improvements in their incomes and conditions. Attempts to repress industrial organisation and then to channel it along non-political and moderate lines nevertheless culminated in the establishment of a militant and effective mineworkers' union. The importance of copper to the colonial economy, the rapid concentration and stabilisation of the workforce and the early and sustained use of industrial action by this workforce to further their interests before and (particularly) after the introduction of trade unions, created awareness of its potential influence amongst nationalist politicians. The ANC and then UNIP sought to ensure that the union movement in general and AMWU in particular ascribed to its agenda of nationalist activity and its vision of post-independence Zambia.

AMWU President Lawrence Katilungu represented the responsible, politically moderate trade union leader sought by colonial officials and the mine companies. His increasing unpopularity amongst his membership and his removal in 1960 indicate that his views were not shared by most mineworkers, many of whom were active in nationalist organisations. Mineworkers took industrial action in defiance of Katilungu's directives, an early sign of the rank-and-file autonomy and organisation that would consistently display itself after Independence. The leaders who replaced Katilungu in AMWU were UNIP members who supported the struggle for national independence. They believed, however, that independence was not simply an end in itself; rather, it provided the basis for organised workers to further their social and economic aspirations in a free and democratic environment.

UNIP's vision was different. Like nationalist organisations elsewhere in Africa, it sought to establish itself as the sole voice of popular aspirations. UNIP therefore tried to sublimate independent unionism, and to control when industrial action occurred and for what purpose. UNIP had gained significant control over much of the Northern Rhodesian African union movement by Independence. However, AMWU's powerful position in the strategic copper mining industry, its organised and active membership and its substantial financial resources frustrated UNIP attempts to control or replace it. AMWU's experience of such attempts provided a salutary warning of what was to follow. The struggle over who would control and benefit from the natural resources and labour concentrated in the mining industry was to be a central political issue in independent Zambia.

2

ZAMBIA'S POLITICAL ECONOMY, 1964 - 1991

Introduction

This chapter analyses the political and economic history of Zambia (and in particular the Copperbelt) from national Independence, until the return to multi-party democracy in 1991. This period was marked by both a dramatic change in economic context in the mid 1970s, but also by an essentially continuous dependency, both economic and political, on the revenue produced by copper mining. In many respects, the primary cleavage in Zambian history was not national Independence, but rather the sudden and sustained collapse in the international copper price in 1974-5 and the related negative shift in Zambia's wider terms of trade. This presaged a decline in the country's economic circumstances and in the livelihoods of ordinary Zambians with virtually no parallel anywhere in the world. Those who determined economic policy, both Zambia's government and its international advisors, nevertheless continued in practice to place the mining industry at the centre of economic policy.

President Kaunda's claims to be moving Zambia 'towards economic independence' were belied by a consistent practical reliance on foreign advisors, investment and (later) aid.[1] Economic policy was made by foreign experts in conjunction with Zambia's Ministers of Finance, a post to which Kaunda generally appointed pro-Western economic liberals. Whilst any significant attempt by Zambia to influence the value of copper would certainly have been met with international opposition, UNIP's general unwillingness (one exception, CIPEC, is described below) to even contemplate such a challenge was not a reflection of simple dependency, but rather one of policy choice, albeit a choice taken (as always) in limited circumstances. The Government was consequently reduced in practice to the redistribution of 'national' wealth amongst sections of the Zambian people. In the long run, the primary task of UNIP administration was the increasingly coercive) management of the expectations of the Zambian people for post-colonial advancement and development.

The post-colonial economy

Zambia achieved Independence in October 1964 in the context of expanding industrial growth in the 'developed' world, based in part on the exploitation of raw materials produced in Africa and elsewhere, such as copper. The 1964 United Nations 'Seers' report envisioned that Zambia's copper revenue would fund substantial growth and development.[2] It implicitly accepted that the copper price would continue to be set on the London Metal Exchange (LME). Whilst acknowledging the need to diversify the economy and overcome copper dependence, the report argued that this diversification would itself be funded by increased copper production. Foreign investment would be needed to develop the mines, necessitating an attractive environment for investment. Seers declared:

> The main interest of the companies (a safe and growing body of profits) is not incompatible with the chief concern of the Government (to be able to rely on an upward trend in revenue to finance its development plans).[3]

This framework, however, effectively limited developmental expenditure to that which could be generated by income from the copper mines. In this context, Seers noted with concern the rapid increase in wages since the mid 1950s and stressed the need to control these in order to channel funds into development:

> The wage and salary question is perhaps the most serious problem facing the Government; its decisions on the wages and salaries it pays itself, and the way it exercises its influence on wages and salaries in the private sector, may be decisive in determining whether Zambia will become during this century a modern developed country…[4]

Economic policy was shaped by this orthodox framework. UNIP's first Government generally followed Seers' strictures in its 1964-66 Transitional Development Plan. Whilst Minister of Finance Arthur Wina emphasised the role of Government in directing economic policy, 69% of the Plan's investment target was expected to come from private sources.[5] Mining companies' incentive to invest was, however, limited by the substantial reduction in gross profits following the Government's eve-of-Independence purchase of mining royalties. Further tax changes meant that by 1966, roughly three-quarters of mine revenue accrued to the state, an amount that quintupled in real terms between 1964 and 1969.[6] Private investment in non-mine sectors was also low, and decreased after the Unilateral Declaration of Independence (UDI) by (Southern) Rhodesia in 1965. UDI, and UNIP's opposition to it, isolated Zambia from a primary regional source of capital and substantially increased export and import costs (Rhodesia's 40% share of Zambian imports in 1964 fell to 5% by 1971).[7]

This failure to attract private investment led many in UNIP to increasingly see the state as the primary engine of economic growth. What remained constant, however, was the utilisation of copper revenue as the primary source of development funds. In 1967, Zambia was instrumental in the creation of the Inter-Governmental Council of Copper Exporting Countries (CIPEC), an attempt by five copper-producing countries to influence the copper market by managing supplies to the market. Self-consciously modelled on OPEC, CIPEC did not, however, control a sufficient part of the copper market to influence the LME price.

Whilst the copper price remained high, significant funds were channelled into loans for emergent farmers to boost production. Substantial social investment fulfilled some expectations of post-Independence development. The state-owned Industrial Development Corporation (INDECO) established new industries producing consumer goods previously supplied from South Africa or Rhodesia, often in joint ventures with private Western companies.[8] Whilst such expenditure helped secure the initial legitimacy of UNIP, it did not provide the basis for self-sustaining economic growth. Agricultural loans were widely regarded as non-refundable rewards for political support. Most INDECO subsidiaries struggled to achieve profitability. Agricultural exports remained stagnant and domestic food production struggled to keep pace with the growing population.

Nevertheless, state expenditure increased in line with the rising copper price and expanding production (copper output rose from 632,000 tonnes in 1964 to 747,000 tonnes in 1969). The Government initially maintained relatively high social spending and patronage of its support base.[9] Whilst it sought to avoid pay increases, it was able to concede them when it became politically unavoidable. In 1969, 51% of the copper mining industry was nationalised with the agreement of AAC and RST, which were handsomely compensated (mine nationalisation and its consequences is examined in Chapter Three). Between 1968 and 1970, much of the Zambian economy was brought under INDECO control, usually with the cooperation of its previously private owners. State ownership did not, however, alter the underlying dynamics of the Zambian economy, failing to reduce the country's dependency on copper.

Post-Independence politics, 1964 - 1972

Zambia's First Republic has generally been understood as a period of relative political harmony in which UNIP's domination of the political sphere reflected its popular support. This, it was argued, stemmed in large part from economic growth and the new Government's consequent capacity to significantly expand the provision of social services and infrastructure.[10] New evidence, particularly from the recently opened UNIP archives, suggests a more conflictual environment.[11] The archives provide overwhelming evidence of UNIP's inability to meet the myriad expectations of post-Independence transformation amongst diverse sections of Zambian society, in all areas of the country. The failure to address these aspirations, themselves originating in the promises made by UNIP leaders seeking to mobilise popular support during the nationalist struggle, led to significant discontent and divisions within the ruling party. Conflicts over the regional allocation of development funds intensified from 1969 as economic stagnation led to cuts in Government spending.

Regional expressions of discontent were heard in virtually every part of Zambia, but took different forms, arising from historical, political and economic particularities.[12] Whilst UNIP secured the electoral support of most Zambians in 1964 and 1968, the consistent opposition of the ANC in Southern Province was reinforced by increasing discontent in Western Province. From 1967, ethno-regional conflicts within UNIP between politicians from Northern and Eastern Province divided the party. Kaunda responded by centralising power within the party and effectively blurring the distinction between the party and the Government by, for example, the appointment of Ministers of State to UNIP's headquarters in 1969.

Bemba politicians from Northern and Copperbelt Provinces expected their numerical dominance in UNIP and their prominence in the nationalist struggle to be reflected in central Government representation and the regional distribution of state expenditure.[13] Kaunda, claiming neutrality but in practice increasingly aligned with the Eastern group, reduced Bemba representation in the Cabinet in 1969. Vice President Simon Kapwepwe, the most prominent Bemba leader, was increasingly marginalised, eventually resigning from Government in 1970. Bemba discontent, combined with growing criticism of corruption and patronage, culminated in the establishment of Kapwepwe's UPP in 1971. UNIP's concern about the threat the UPP offered to its traditional support base prompted the declaration of the one-party state in December 1972.

A major political preoccupation was the post-Independence influx of Zambians into towns. The removal of colonial population controls and the lack of rural development led to a major growth in the urban population and the proliferation of unofficial settlements, at a time when few productive jobs were created. Advisors warned of the dangers of an expectant new urban population unable to secure jobs or economic opportunities. Kaunda's Humanist ideology suggested that the authentic Zambia was that of villagers and agricultural production. Consistent appeals were made to go 'Back to the Land', and a range of initiatives, such as the Zambia Youth Service, were introduced to provide skills training in agriculture. Whilst this replicated colonial anxiety about potential urban unrest, UNIP was unable and/or unwilling to impose influx controls.

Whilst UNIP failed to achieve rural development, the rhetoric surrounding it formed a central theme in attempts to control urban wages. This was based on the developmentalist assumption, unsupported by evidence, that wages not taken in urban areas would somehow find their way into productive rural investment. This approached disregarded factors relevant to the workers involved: that urban wages remitted to relatives provided a substantial source of investment in commercial ventures in rural areas; and that higher wages enabled workers to support increased numbers of relatives to live in towns whilst they received education, looked for work, or developed informal enterprises.

Economic control and dependency, 1973 - 1978

The introduction of the one-party state was followed by international economic developments over which UNIP had no control, but which had a decisive impact on Zambia's future. The rise in global oil prices from 1973 and the subsequent worldwide recession led to a severe and lasting decline in international metal prices. Having risen steadily during the 1960s, the copper price briefly peaked at £1,400 per ton in 1974, then collapsed to £500-£600 per ton in 1975 (see Appendix Two). In real terms, the copper price never recovered. This destroyed the assumptions on which Zambia's developmental path had been constructed. Rhodesia's closure of the border in 1973, subsequently enforced by Zambia until 1978, added to the cost of exporting copper. The Angolan civil war and the consequent closure of the Benguela railway in 1975 was a devastating blow to the economy, not significantly alleviated by the opening of the

TAZARA route to Dar es Salaam in 1976. Private manufacturing companies closed, or reduced production: 20,000 jobs were lost in 1976 alone. Foreign exchange reserves fell from K76m in 1975 to K12m in 1976.[14]

Zambia substituted for lost copper revenue by international borrowing, initially via the burgeoning Eurodollar market.[15] When these debts could not be repaid, Zambia became dependent on Western donors and the IMF, with which an agreement was first signed in 1973. Job creation and economic diversification envisaged in (increasingly unrealistic) development plans were not achieved. Goods produced by the parastatal sector, commonly assembled from foreign inputs, were (because of the rise in input costs and the devaluation of the Kwacha) now more expensive than imports.

Whilst UNIP rhetoric continued to emphasise economic self-reliance, the effects of the falling copper price starkly revealed Zambia's inability to reduce its dependency on the international economy, into which it was integrated solely as a supplier of copper. Attempts were made to compensate for this lack of control with an increasingly interventionist economic policy. Kaunda unexpectedly redeemed the bonds held by the international mining companies in 1973, repudiating their contracts for management and marketing and establishing the state-owned Metal Marketing Corporation (MEMACO). This was portrayed as a radical step towards local control, but it may have resulted from links between UNIP and international business interests.[16] Redemption certainly did not reduce Zambia's dependency on external finance, since it required additional foreign borrowing from commercial lenders. Minister of Mines Andrew Kashita assured the LME that redemption would have no impact on their commercial relationship:

> Arrangements were to be made so that senior staff of the present metal marketing Companies in London would in future work for Memaco and customers would find that they were largely dealing with the same people, "but with a different label over their heads."[17]

In 1974, the mine companies appointed their first Zambian Managing Directors. Effective control, however, remained with expatriate General Managers, who ran each mine with significant autonomy from senior management. The decline in foreign exchange earnings severely hampered the mines' capacity to purchase inputs, contributing to a fall in mine production during the 1970s (see Appendix Two). The companies also accumulated large and, in retrospect, unpayable debts. In 1978, the Government's increase in its shareholding in the mine companies to 60.7% effectively represented a state loan.

UNIP's response to economic decline fluctuated between increased state control and limited market reforms, the latter driven by the increasing influence of the IMF over the indebted Zambian economy. Despite consistent appeals for Zambians to return to the land, agricultural development remained poorly funded, contributing less than 10% of Gross Domestic Product.[18] The failure of rural development encouraged continued urban migration; the population of the Copperbelt tripled between 1964 and 1979.[19] Urban unemployment was periodically addressed by policies such as rural reconstruction centres, where the jobless were to be trained in agricultural skills. Such policies, haphazardly implemented, did not significantly address these problems. Policy incoherence also reflected tension within UNIP. Older nationalist leaders, who sought to centralise economic power in the state, clashed with a younger business-oriented group, many of whom had acquired substantial wealth through state positions and who sought to realise its value in a freer economic environment. Conflict over the emergence of a business elite was reflected in the contested implementation of UNIP's Leadership Code, meant to restrict the personal wealth of Party leaders.[20]

As incomes failed to keep pace with the rising cost of living (official inflation rose to 35% in 1975 and to 50% in 1976), the Government addressed demands for higher wages by increasing subsidies on essential goods.[21] Whilst this controlled prices, it also led to widespread smuggling and consequent shortages. Access to essential goods, particularly the staple mealie meal (maize meal), became increasingly dependent on one's economic position and links to UNIP, contributing to a sense of inequality and class discontent amongst unionised workers. In 1977, a Parliamentary Select Committee report chaired by Minister of Finance John Mwanakatwe recommended more market-oriented economic policies.[22] Many of its findings were rejected and UNIP's strongest pro-market advocates were marginalised. Government policy nevertheless moved significantly in this direction, for example in directing parastatals to charge 'economic' prices. In 1978, Kaunda opened the Rhodesian border. In 1979, directives were issued to close loss-making companies.

In 1978, Zambia signed a new agreement with the IMF, agreeing to reduce food subsidies and devalue the Kwacha by 10%. The agreement imposed severe credit restrictions on the mining industry.[23] Borrowing was restricted and total Government expenditure was slashed from

K150m in 1977 to K90m in 1978.[24] IMF loan payments were made conditional on the setting of mine production targets and on the continuation of the 5% annual wage increase limit, placing the Fund in indirect conflict with the labour movement.[25] The decline in the international copper price and its devastating impact on the national economy meant that UNIP was unable to make concessions to workers. When implementing unpopular economic polices, it became increasingly important to secure the support, or neutralise the opposition, of the labour movement.[26]

Regional conflict and political repression

A primary characteristic of UNIP governance was the utilisation and magnification of the threat that it faced from its regional enemies, in order to justify its repression of internal dissent.[27] In the late 1960s, Rhodesian UDI provided a justification for an increasingly centralised and authoritarian administration. Kaunda justified banning the UPP on the (entirely unproven grounds that it was receiving military support from Zambia's enemies.[28]

Zambia's hosting of regional liberation movements led to periodic attacks by Rhodesian and South African forces. In the early 1970s, attacks were launched against Zambia from Portuguese-ruled Angola and Mozambique. At the same time, South Africa remained an important trading partner of Zambia throughout the 1970s and 1980s, with AAC retaining a significant minority share in Zambia's primary industry. Despite public hostility between the countries, indirect diplomatic links were maintained. In the mid 1970s in particular, South African and Zambian political representatives cooperated closely in promoting political settlement in Rhodesia and in supporting UNITA in Angola.[29] UNIP had little choice but to maintain necessary ties with its hostile neighbours, but reserved the exclusive right to define what links were acceptable, habitually defining any challenge to its judgement as treasonous. In 1978 for example, Kapwepwe was demonised for proposing the reopening of the Rhodesian rail link. Months later, Kaunda justified his decision to reopen the railway (as Kapwepwe had) in terms of economic necessity.

As the economic crisis worsened, UNIP's use of radical rhetoric to disguise its true economic orientation grew. A year after Zambia signed a new austerity agreement with the IMF in 1978 (see below), Kaunda announced 'war measures', claiming that western Governments and international capitalism were in league with racist regimes to overthrow African 'socialist' countries including Zambia.[30] Zimbabwean independence in 1980 reduced the political justification for emergency measures. The police nevertheless increased the use of mass arrests, detentions and beatings, ostensibly to combat crime.[31] District Governors used sweeping local powers to, for example, ban women from queuing at State shops for essential commodities, a policy that was enforced with beatings by UNIP 'militants' and vigilantes. In 1980, Kaunda linked the trade union movement to a coup plot, without presenting any evidence for such a connection.

Political relations under the one-party state, 1973 - 1980

The Second Republican Constitution legally entrenched UNIP supremacy and accelerated the shift of power from ministries and the Cabinet to the Central Committee and its sub-Committees. UNIP's political dominance was however, like its control of the economy, more official than effective. On the Copperbelt in particular, the legacy of the conflict with the UPP left UNIP weakened by a haemorrhage of membership and a resultant low turnout in subsequent one-party elections. As Gertzel argued:

> on the Copperbelt … the 1973 elections had indicated a significant loss of support for U.N.I.P. … the province remained under-represented at the centre … In the late 'seventies, therefore, the potential for political dissent on the Copperbelt went beyond the miners and involved urban labour as a whole.[32]

Attempts to revive the UPP and other forms of opposition (often within UNIP) on the Copperbelt resulted in periodic expulsions of UNIP members. The ongoing fear of a revived opposition led UNIP to restrict its membership criteria in 1974.[33] By 1976, UNIP's Copperbelt membership fell to 32,194 in a population of 1,185,000, at a time when 112,565 members were required to fill all positions in local Party structures.[34] UNIP officials across Zambia reported sharp declines in membership, activity and functioning local party structures.[35]

Partly as a response to the crisis in membership, UNIP sought reconciliation with the UPP's former leaders. Kapwepwe and others rejoined UNIP in 1977. UPP activists, who had maintained secret structures, now sought to nominate Kapwepwe as a UNIP Presidential candidate.[36] In January 1978, prominent former UPP leader Musonda Chambeshi won a by-election in Roan mine township.[37] Chambeshi attacked the high cost of living that, he claimed, put workers on a collision course with the Government.[38] Following accusations of plotting against Kaunda, he was suspended from UNIP and lost his Parliamentary seat.[39]

Kapwepwe presented a threat sufficient for UNIP to amend its Constitution to prevent any challenge to Kaunda, who was elected unopposed in 1978. The Copperbelt was wracked by tension and violence, with anonymous 'UPP' circulars being denounced by UNIP. Internal UNIP documents identified the threat posed by former UPP members and recommended propaganda campaigns against the banned party.[40] During this period, MUZ leader David Mwila was briefly arrested, accused with his union of links to the UPP (see Chapter Four). Former UPP leaders were detained and severely tortured.[41] The last of these were only released in 1981. Kapwepwe's challenge was ended by his death in 1980.

Economic decline in the 1980s

The 1980s saw an acceleration of the decline in the Zambian economy. There were brief periods when the international copper price rose and a degree of recovery seemed possible, but the dominant pattern was one of increasing economic failure and rising poverty. Per capita income fell from $630 in 1981 to $200 in 1987 and increasingly thereafter.[42] Government income fell by 30% between 1974 and 1984.[43] A 1986 World Bank report found that most Zambians were poorer than they had been in 1970. Formal sector employment fell from 381,490 in 1980 to 359,620 in 1989.[44] By 1984, Zambia's national debt had reached $4.5bn.[45] In response to declining employment opportunities, the Copperbelt population began to fall during the 1980s, with moderate out-migration to other areas.[46]

Whilst official inflation rates in the early 1980s averaged approximately 15-20%, continuing shortages drove up black market prices to three times their official price by 1982.[47] By the mid 1980s, the black market exchange rate between the US dollar and the Kwacha was four times the official rate.[48] Shortages of foreign exchange led to low productivity in industry and transport: in 1985, 320 of 555 state-owned buses were not running because of a lack of spare parts.[49] Charges were reintroduced for health and education services.[50] Public awareness of economic mismanagement and corruption in the Government and parastatals grew. An increasing number of corruption cases came to public attention, for example the Tika Steel project, which resulted in the illegal transfer of large amounts of money to senior politicians.[51] Criticism of the lack of essential commodities mounted, as did attacks on the *apanwamba*, the wealthy political elite who had never suffered hunger and whose mismanagement had put Zambia into debt.[52]

IMF and World Bank influence, 1982 - 1987

The merger of the existing mine companies in 1982 was largely driven by Zambia's donors; the IMF provided major loans for the new ZCCM.[53] Whilst ZCCM was theoretically established to increase efficiency in the exploitation of Zambia's primary resource, in practice it created greater opportunities for the political manipulation of mining economies.[54] ZCCM management was unable to retain sufficient revenue to re-invest. With copper prices remaining low and production costs rising, ZCCM made no substantial profits (and frequent losses) during the 1980s.[55] Copper production stagnated, falling to 525,811 tonnes in 1985.[56] Despite this, UNIP and donors alike continued to emphasise the importance of the company to achieving national economic diversification.[57] ZCCM, unlike the Government, could still borrow foreign currency. As Aron describes, this provided a line of credit for both politically prestigious projects and for luxury jets and holiday resorts used by senior politicians.[58] By February 1987, ZCCM had debts of $800m.[59]

Despite its unprofitability, the enduring centrality of ZCCM to the economy was consistently confirmed. A briefing on its 1986–90 Production and Investment Plan stated:

> The Company's recovery is of critical importance to the success of the current effort by the Party and its Government to restructure and diversify the economy. This is because ZCCM earns virtually all the foreign exchange required in the national economic recovery and diversification programmes.[60]

By 1984, debt servicing consumed 65% of foreign exchange earnings, making Zambia the most heavily indebted country in sub-Saharan Africa.[61] Donor dependence increased the influence of the IMF over Government policy. In the early 1980s, food subsides were cut and the Kwacha devalued by 40%. The IMF demanded the restriction of wage rises to 5%, whilst companies were encouraged to charge 'economic prices'. Kaunda claimed that such conditions would instil a sense of responsibility.[62]

In 1985, an extension of the existing IMF agreement increased conditionality, with the Kwacha being decontrolled and foreign exchange allocated via an auction system.[63] Kaunda called on labour leaders to educate their members regarding the nation's hardships to maintain industrial peace.[64] The Kwacha initially stabilised at K6 to the US$, half of its previous value.[65] The prices of most goods were decontrolled and inflation soared.[66] The ZCTU called for a wage rise, to be fixed against the US dollar.[67] The auction system enabled the legal importation of luxury goods, increasing public awareness of a wealthy state-based elite.[68] Indigenous private business representatives claimed the auction system was biased towards international capital and called for it to be scrapped.[69] The Bank of Zambia subsequently conceded that 99% of the system's beneficiaries were foreign multinational companies.[70]

In December 1986, implement of the IMF agreement led to a doubling in the mealie meal price and to widespread rioting on the Copperbelt (see Chapter Five). The riots forced UNIP to reverse the price rises. Donors responded by reducing financial support, making the auction system unworkable. Major strikes now took place, as workers demanded pay increases to compensate for inflation.[71] Kaunda claimed that 'the initiators of the strikes were politically motivated.'[72] He accused the ZCTU of receiving funds from CIA-backed organisations; Congress in turn warned the Government against scapegoating the labour movement for the economic crisis.[73]

In February 1987, Prime Minister Kebby Musokotwane claimed that the Government retained the capacity to direct economic policy. Such a statement had potentially dangerous consequences: if UNIP was the powerful and effective force it claimed to be, it had to take responsibility for unpopular policies; if responsibility lay with external agencies, the weakness of the one-party state was exposed. Certainly, the accelerating economic decline could not be allowed to continue; the Kwacha fell to K21, one-tenth of its value before the auction. Ultimately, popular discontent proved more influential than the IMF. On May Day 1987, Kaunda announced the scrapping of the auction system; the Kwacha was fixed at K8 to $1, and the price of essential goods was frozen. The World Bank concluded:

> The early demise...of the adjustment package imposed by the IMF resulted from an unrealistic ... assumption that the majority of middle and lower income urban Zambians would tolerate pauperisation...[74]

A New Economic Recovery Programme (NERP) was now introduced, supposedly based on growth from Zambia's own resources. Debt servicing was limited to 10% of export earnings. Donors, however, refused to fund NERP. Despite official rhetorical commitment to popular control over the economy, the withdrawal of donor support meant that, for example, District Councils were unable to pay wages. The re-imposition of price controls meant renewed shortages; inflation remained high, reaching 47% in September 1987.[75]

Despite claims that Zambia would not return to the IMF, UNIP continued to implement market reforms.[76] Following the 1988 election, the Kwacha was devalued by 25%.[77] INDECO signed management contracts with foreign firms to run parastatals.[78] In July 1989, the prices of all goods except mealie meal were decontrolled and the Kwacha was devalued to K16 to the US$. Within days, many prices had doubled: bread rose from K7 to K12.[79] At the same time, Kaunda introduced a 50% wage rise for low income public sector workers, with other workers told to bargain for similar increases.[80] Protests nevertheless broke out on the Copperbelt; riot police guarded state-owned shops in Kitwe's mine townships.[81]

In November 1989, the IMF and World Bank agreed a new economic policy framework.[82] The government pledged to remove all food subsidies, introducing food coupons to subsidise the poorest.[83] The economy nevertheless continued to decline: the Kwacha fell in January 1990 to K23.60 to the US$.[84] Inflation reached 122% by December 1989.[85] In February, Zambia and the IFIs agreed a new austerity programme, with further cuts in public spending. By this time, Zambia's debt had reached $7bn.[86]

Dissent and opposition in the 1980s

Despite periodic military attacks on Zambia by South Africa during the 1980s, the country was also Zambia's second biggest trading partner, providing 16% of imports in 1982.[87] In that year, President Kaunda sought rapprochement with the South African leadership.[88] By the mid 1980s however, Kaunda was again claiming that South Africa was a military threat. In 1987, Kaunda alleged that South Africa had unleashed 'Zambian dissidents' to cause chaos in Zambia.[89] Such claims were not always false, but they provided an increasingly less convincing justification for internal repression of dissent.

Leading politicians periodically proposed a constitutional amendment to permit the forcible deportation of the unemployed to their villages.[90] The demolition of squatter areas, publicly justified as an anti-crime initiative, was frequently motivated by the need to clear prime urban land for speculative development. In 1988, the Minister of Defence began the compulsory rural resettlement of an initial 10,000 youths from the Copperbelt and Lusaka, although this was not fully implemented due to resistance from the target group.[91] The police, reinforced by armed forces and legalised vigilantes, engaged in periodic crackdowns against illegal foreign residents. Vigilantes attacked alleged black marketers; they policed queues for essential goods and ensured preferential access for prominent UNIP officials. The police habitually used violence and sometimes torture against suspects.[92]

UNIP achieved improved election results in 1983 and 1988, with an increased turnout and a higher percentage of 'Yes' votes for Kaunda. The Party, however, continued to struggle with its own recruitment and organisation. UNIP's campaign for 500,000 members in the Copperbelt in 1982 led to the recruitment of just 92,000, in a population of 1.3 million.[93] Despite the centralisation of Parliamentary candidate selection to weed out dissidents, backbench MPs became increasingly critical of Government policies.[94] Business representatives criticised economic policy. In 1989, the Zambia National Chamber of Commerce and Industry (ZNCCI) attacked an economic policy that marginalised indigenous business and favoured multi-national corporations linked to parastatals.[95]

State control of the media increased with the takeover by UNIP of the *Times of Zambia* in 1982.[96] Organised religion, particularly the Catholic Church, criticised human rights violations and established independent media through which opposition voices began to be expressed. Zambians demonstrated their discontent through increasingly frequent food-related riots and localised disturbances. The University of Zambia was closed on at least three occasions during the 1980s following expressions of student opposition to Government policies.[97]

Following a coup attempt in 1988, the trial of its organisers heard evidence from a state witness that Kaunda possessed K4bn in external bank accounts, feeding popular perceptions of the avarice of leading politicians.[98] Kaunda sought to offset the threat of a coup by the increased representation of the armed forces in the Government, symbolised by the appointment of General Malimba Masheke as Prime Minister in March 1989.[99] More generally, UNIP sought to address rising discontent through the incorporation of potentially critical voices into its structures. In 1988, the UNIP Central Committee was expanded from 25 to 68 members, to include representatives of the military, parastatals, and the Mineworkers' Union of Zambia (see Chapter Five).[100]

The Movement for Multi-Party Democracy, 1989 - 1991

The overthrow of East European one-party states in late 1989 had a significant impact in Zambia. On 31 December, ZCTU National

Chairman Frederick Chiluba became the first prominent Zambian to publicly declare that African countries including Zambia should consider the re-introduction of multi-party systems.[101] In May 1990, Kaunda conceded a referendum on multi-partyism, but argued that such a system would re-ignite tribalism.[102] Meanwhile, Kaunda announced that 49% of shares in some parastatals would be sold and a stock exchange created.[103] Kaunda, however, failed to resist donor pressure to remove food subsidies. In June, mealie meal prices more than doubled.[104] In Lusaka, thousands looted and rioted before a curfew was imposed – 36 people died.[105] Subsequently, an Army Lieutenant launched an opportunistic coup attempt. Whilst this was unsuccessful, it was initially celebrated on the streets. By publicly revealing popular opposition to UNIP, these events provided a 'moment of transparency' and a significant spur to the emergent campaign for multi-party democracy.[106]

In July 1990, multi-party advocates met in Lusaka.[107] As well as Chiluba, these included former UNIP leaders, intellectuals, lawyers and prominent businessmen. In August, the unregistered multi-party democracy movement held its first rallies in Copperbelt towns.[108] Such events, enabled by the mobilisation of trade union resources, helped prevent the suppression of the multi-party campaign during its period of ambiguous legal status. Kaunda initially postponed the planned referendum and then declared in September that Zambia would hold multi-party elections in August 1991.[109] UNIP initially sought to prevent multi-party advocates from holding meetings. Permits were denied; Chiluba was arrested at a rally and charged with unlawful assembly.[110] The opposition warned that such actions might lead to a break down in law and order.[111] In practice, the overwhelming public support for the multi-party movement deterred UNIP from any systematic attempt at its repression. Kaunda subsequently confirmed that he abandoned plans to detain Chiluba for fear of the consequent unrest.[112]

In December, the multi-party movement re-launched itself as a political party, the MMD.[113] In March 1991, Chiluba was elected MMD President.[114] Whilst the unity demonstrated by multi-party advocates enabled a single coherent challenge to UNIP, there was consequently little public discussion of its policies. MMD interim chairman Arthur Wina privately argued in 1991 that Chiluba was elected party President because, with his trade union background, he was the only leader who could implement economic adjustment policies without the opposition of organised labour.[115] In November 1990, Wina assured World Bank officials that an MMD Government would honour Zambia's debt obligations.[116] The MMD manifesto proposed a 'social market economy', with Government as a facilitator rather than a participant in the economy.[117]

In May 1991, radically reversing his previous position in ZCTU (see Chapters Four and Five), Chiluba declared that privatisation would play a major role in the economy.[118] An MMD advert proclaimed, 'Across the world, authoritarian governments and centralised economies are on the run: While democracies and free markets are on the rise. Economic competition will be unprecedented and so will economic opportunities.'[119] UNIP initially attacked these policies as a copy of their own privatisation policies.[120] The market reforms promised by both parties reflected popular discontent with the deteriorating economic situation. Prices more than doubled during 1991 and food shortages continued. Donors punished UNIP's refusal to raise the mealie meal price by withholding funds.[121]

The October 1991 election was generally peaceful. Kaunda quickly conceded defeat; Chiluba received 75% of the vote for the Presidency. The MMD won 125 Parliamentary seats, to UNIP's twenty-five.[122] Days later, paramilitary police raided ZCCM offices.[123] Investigations revealed that 25 ZCCM subsidiary companies had been established outside Zambia without legal approval. Even in the final days of UNIP rule, ZCCM remained central to the ruling party's mode of governance, which by this time was transparently based on the extraction of national resources for personal and political gain.

Conclusion

This chapter has sought to demonstrate the enduring dependency of Zambia's economy and society on the copper mining industry. Notwithstanding the pledges of UNIP leaders at national and local level to transform economy and society for the benefit of ordinary Zambians, UNIP's macro-economic policies ensured that the funds for this transformation were perpetually restricted to that which could be funded from income from international copper sales (the value of which was defined in markets beyond their control). This necessitated an accommodative relationship with the international mining companies, a relationship that, political declarations and changes in mine ownership notwithstanding, remained qualitatively unchanged during the period under examination.

Within this framework, UNIP's primary task was to manage, rather than to address, the expectations held by Zambians for Independence. There was always a gap between these aspirations and what the post-colonial state was able to deliver; that gap grew steadily and then dramatically with the decline in copper prices from the mid-1970s. From this point on, the state's central task was to police the discontent arising from the resultant decline in living standards. In the long-term, the resistance of disparate sections of Zambian society to this decline provided the basis for the pro-democracy movement in 1990-91. Subsequent chapters will explore the efforts of one section of that society, Zambia's mineworkers, to overcome the UNIP-dominated state's resistance to the realisation of their expectations of post-colonial economic and social transformation.

3

FROM INDEPENDENCE TO THE ONE-PARTY STATE, 1964 - 1972

Introduction

In contrast to a widespread acknowledgement of the importance of civil society activity in favour of political and economic change in the 1980s and early 1990s, there is a common assumption that mass organisations offered little challenge to post-colonial states in the 1960s. This is generally thought to be because, during a 'honeymoon period' of post-Independence economic growth, there was a relative consensus over state-led development. It was only with economic stagnation from the early 1970s that significant popular disillusionment with the failure of the post-colonial state to deliver arose. Certainly, as subsequent chapters will show, economic decline undoubtedly decreased the capacity of African states to meet their peoples' expectations, intensifying social conflict and undermining democracy. However, this chapter argues that from Independence in 1964, the failure of UNIP to address the expectations of many Zambians led to significant domestic opposition and social conflict.[1]

Where the existence of such conflict is acknowledged, it tends to be effectively de-legitimised in terms such as Schraeder's 'great expectations – minimal capabilities paradox':

> The newly elected leaders had to contend with popular expectations that the fruits of independence, most notably higher wages and better living conditions, would be quickly and widely shared ... however, the former colonial state did not have the capabilities to satisfy public demands ... The vast majority of African leaders resolved this paradox by promoting the concentration of state power at the expense of civil society.[2]

Such an analysis disregards the origins of such expectations in the promises of nationalist politicians that were necessary to mobilise popular opposition to colonial rule. Such promises were implicitly predicated on the radical redistribution of wealth, both nationally and internationally. The failure of post-colonial governments to seek such redistribution was not a technical inevitability but the result of political decisions that, as this chapter will demonstrate, were actively contested by popular movements.

Bates understood that controlling the expectations of Zambia's copper mineworkers was a central problem for the post-colonial state, arguing that, 'Given the country's reliance on copper and its aspirations for rapid economic growth, the control of the behaviour of the copper miners represents a crucial political task…'. [3] Bates described UNIP's attempts to turn the union movement into 'productionist' organisations. However, his effective conflation of party and national interests led him to overstate the extent of mineworkers' support for UNIP. By equating union effectiveness with ability to control members, Bates discounted their importance in addressing members' aspirations.[4] Such an analysis is emblematic of the way in which the diverse expectations of heterogeneous African societies were subsumed under a single nationalist meta-narrative, tacitly defining many expectations as illegitimate.[5]

Much of the established literature on political parties and labour organisations in 1960s Africa claims that, because the economic gains from Independence were channelled primarily to urban areas, organised labour did not mobilise on class terms. Tordoff claims:

> Urban workers in many countries benefited from food and other subsidies (often at the expense of the rural population) … Upward mobility, or the prospect of it, dampened class polarisation even when differences of status and wealth were clearly perceived...[6]

Such assumptions are doubly misleading. Firstly, African formal sector wages in all main sectors, having risen substantially up to and during the immediate period of Independence, declined in real terms in the subsequent decade. This decline in real wages was not entirely alleviated by new post-colonial subsidies; the living standards of most organised workers were at best stagnant and in many cases in decline during this period. Secondly, the impact of the 'perceived upward mobility' claimed by Tordoff depends on the specific expectations workers had for post-Independence improvements in their working and living conditions. If these were substantially greater than what was actually delivered, the effect would presumably be increased, rather than dampened, class polarisation.

Specific analysis of this period in Zambia is generally confined to electoral politics and the official policies of institutions such as trade unions. For example, Nordlund focuses on apparent good relations between union *leaders* and UNIP:

> Another indication of good government and union relations (at least at leadership levels) during the first years after independence is illustrated by the fact that a number of trade union leaders left their unions to work actively within UNIP. Both sides felt to a large extent that they were working towards the same goal of national development...[7]

Trade unionists appointed to Government were in practice criticised for seeking to serve the contrasting interests of workers and Government. Ministers consistently attacked the inability of union leaders to prevent unofficial strikes. Precisely because 'good relations' did not provide UNIP with effective control over unions and their members, the ruling party sought via legislation to create a labour movement, and a mineworkers' union in particular, that would contain workers' particular definition of 'national development'.

Despite noting Burawoy's finding that in the early 1970s, 'mineworkers on the Zambian Copperbelt saw their own union leaders as part of a new Zambian elite', Tordoff fails to identify the significance of *intra*-union conflict, or to understand that the discontent of organised workers with the post-colonial economic settlement was, as this chapter will show, actively expressed in non-institutional forms. Burawoy certainly identified mineworkers' sense of injustice with what they regarded as the slow pace of post-colonial change. He noted that in post-Independence Zambia,

> The black ruling-class has taken over many facets of the colonial ideology to buttress and legitimize its own rule ... like other managerial or ruling-class ideologies, [it] proclaims the worker as indolent and undisciplined.[8]

Burawoy's analysis is, however, partly vitiated by his characterisation of the mineworkers as a 'labour aristocracy', no evidence of which is provided in his interviews.[9] Burawoy suggests that 'The Zambian worker['s]... interests as a worker clash with his interests as a Zambian.'[10]

The evidence presented in this chapter indicates that mineworkers saw no contradiction between these interests. To achieve their vision of an independent Zambia in which its peoples would achieve rapid political and economic advancement, they sought to extend independent local union organisation and to utilise industrial action. Major strikes in 1966 demonstrated that the improved wages and conditions they saw as a reward for their prominent role in the nationalist struggle would (as under colonialism) only be achieved through industrial action in the face of opposition from company, Government and union officials.

Mineworkers sought to elect union leaders whose autonomy from UNIP would, they believed, enable them to achieve improvements in wages and conditions. This culminated in 1971 in a challenge to their union leadership that was ended only by its suppression by the state. Mineworkers' electoral support for UNIP was contingent on their expectation of advancement. When the limits of advancement became evident, a significant part of the mine labour force supported a breakaway opposition party. Throughout this period, the conflictual relationship between the union bureaucracy and its membership lay at the heart of mineworkers' experience of post-colonial political and economic change.

Government and unions at Independence

The centrality of wage control (see Chapter Two) made, in turn, the control of labour a major issue for post-colonial governance. As in the years before Independence, UNIP was preoccupied with achieving control over Zambia's labour movement and the mineworkers' union in particular. Party leaders advocated a new structure of Zambian unions, with responsible trained officials insulated from foreign influences. Kaunda claimed many union leaders were stooges of foreign powers and they were frequently prevented from attending overseas training courses.[11]

The importance of this issue can be gauged by the publication, weeks after Independence, of a Bill to establish a 'Zambia Congress of Trade Unions' with significant powers to create and dissolve member unions, alter their rules and to approve or reject strike ballots and industrial action.[12] Minister of Labour Justin Chimba denied it was Government's desire to control trade unions.[13] However, his briefing to a Cabinet Committee reveals the reality:

> … the British Government did not agree to Government's request for an amendment to the new Constitution, which would have allowed the enactment of a more directly restrictive Bill. The Bill …

> is therefore designed to achieve the maximum possible control over trade unions and their officials largely by indirect means ... It is hoped that its general effect will be to encourage the development of a single union in respect of each trade, industry or occupation, and, by making membership of the Zambia Congress of Trade Unions attractive to such unions, facilitate greater control over their policies by a responsible body which, though not subject to Government control, will assist in giving effect to Government policy...[14]

The resulting Trade Unions and Trade Disputes (Amendment) Act was passed in January 1965. The Act provided the Ministry of Labour with extensive powers over the new ZCTU, including the right to appoint its first officials and the power to approve or withhold approval of amendments to its rules.[15]

The existing United Trade Union Congress (UTUC), only established in June 1964 under the guidance of the Ministry of Labour (see Chapter One), initially resisted dissolution to make way for the new ZCTU. However, its lack of both funding and popular legitimacy limited its capacity to resist Government decisions.[16] Many veteran unionists were drawn into Government service, further weakening union structures. MUZ President John Chisata, for example, was elected as MP for Mufulira. These were also undermined by the removal, under the 1965 Act, of eleven trade union leaders who had never worked in the industries they represented, a decision deplored by the UTUC.[17] The UTUC only agreed to its dissolution following an ultimatum:[18]

> [UTUC President] Comrade Mugala ... did not approve the system of appointing officials to Zambia Congress of Trade Unions, but appealed to the Government to call a wonderful Conference where all intended leaders could be elected democratically. He further maintained that, if only these will be appointed it follows therefore that they will be Government stogges [sic - stooges]; because any one so appointed will automatically become a civil servant ... Since the introduction of Trade Union Act, the labour movement have not enjoyed the freedom we used to have, namely Freedom of Movement.[19]

Nevertheless, in December 1965, the Ministry of Labour appointed Augustine Nkumbula as the first ZCTU General Secretary; no UTUC officers were appointed to the new body.[20] Minister of Labour Nalumnio Mundia claimed that the new officials were representative of the labour movement.[21] Nkumbula asserted this was not a Government takeover 'because they [the Government] recognise the importance of having a well organised labour movement that would contribute to nation building as a whole.'[22]

A sceptical Zambia Mineworkers' Union (ZMU) delayed affiliation to the ZCTU. As the largest and wealthiest union, the ZMU's affiliation was essential not only for ZCTU sustainability, but also for Congress to present itself as the legitimate voice of Zambian workers. Nkumbula complained of the lack of funding and privately asked Government to fund Congress until the check-off system had been extended.[23] Member unions attacked ZCTU inactivity; little of the proposed training of 'responsible' new union officials took place. Ministry of Labour officials criticised Congress for consistent unauthorised expenditure.[24] Frustrated, Nkumbula resigned in September 1966.

The ZMU affiliated to the ZCTU in June 1966, but expressed its lack of confidence in the Congress leadership, repeatedly requesting their direct election. Ministry of Labour officials privately admitted this could no longer be resisted:

> … the Congress of Trade Unions can not enjoy the confidence of member trade unions if we were to continue appointing its officers. This has been expressed by so many Union officers such that their contention is that, Z.C.T.U. officials have to act on instructions of the Government in Lusaka rather than on the interests of member unions or workers … It may be argued that once these elections are allowed we may have elected officials who may be impossible to cooperate with. This might be true but can be remedied by extending the present powers conferes [sic] on the Registrar of Trade Unions to cover the Z.C.T.U. … Finally, I should suggest that to have irresponsible trade unions leaders under control, the powers to disqualify certain trade union leaders should be extended … our Country is young and as such, trade unions must be controlled.[25]

A new leadership was elected in July 1967, but the Ministry retained an effective veto over candidates. Wilson Chakulya, the new General Secretary, expanded the ZCTU's training programme, supported by the International Labour Organisation (ILO).[26] Congress expansion was enabled by ZMU/MUZ subscriptions, as many other unions failed to make contributions. In 1969, MUZ accounted for 40% of the ZCTU's membership of 115,379 and contributed 61% of ZCTU income in 1970.[27] Congress admitted, 'MUZ is the only up to date paid up union to this Congress. Without this Union and few others the Congress would not have been in existence because financially we depend on them.'[28] Chalukya's requests to improve Congress finances by increased controls over member unions were, however, rejected.[29] The Registrar of Trade Unions (RTU), opposing such changes, declared that 'It is my understanding that the Government's intention is to keep the Congress small so that the Minister can easily and effectively exercise control over this body.'[30] The debate over whether the ZCTU, or the Ministry of Labour, should take the lead in controlling unions and their members, remained a hotly contested issue (see below).

Under Chakulya, Congress positioned itself as the voice of worker loyalty to UNIP. Chakulya not only called on members to vote for UNIP in the 1968 elections, but also for opposition parties to be banned.[31] The ZCTU organised Zambia's first May Day demonstrations that year; their tone can be judged from the report that, on May Day 1969, 'Marchers will display placards condemning wildcat strikes; calling for discipline, increased productivity, and unity between the ZCTU and UNIP, and thanking President Kaunda for worker participation in industry.'[32] Member unions, however, complained about increasing Congress interference in their affairs. Attempts to remove Chakulya in 1968 were prevented by the closure of a ZCTU conference before a vote could be taken.[33] Chakulya's election to Parliament that year led to calls for his removal; criticism of his dual role led to him being labelled as 'Government's boy'.[34] He was nevertheless re-elected in 1970, amidst condemnations of the undemocratic selection of and pressure on the delegates electing him.[35]

The ZMU and UNIP, 1964 - 1965

In the wake of Independence, Zambia's copper mines were affected by hundreds of local disputes as mineworkers sought to challenge the racial and violent system of supervision on which production had rested since the 1930s. Miners practically undermined discrimination by, for example, occupying changing rooms reserved for white mineworkers. Branch officials sought to extend their influence following a new recognition agreement signed in October 1964, which for the first time permitted shop stewards at all workplaces.[36] Some General Managers, however, resisted the appointment of stewards and undermined their capacity to work effectively. In seeking to turn these paper gains into practical ones, mineworkers expected support from the new UNIP Government. However, whilst local party leaders frequently involved themselves in local disputes, particularly those concerning racial discrimination, national UNIP leaders consistently condemned strikes.[37]

Cypriano Musonda was dismissed from RST's Ndola Copper Refinery (NCR) days after Independence for allegedly refusing to carry out the order of an expatriate supervisor. A local protest strike was called off by branch officials after two days when ZMU leaders intervened.[38] RST denied that Musonda's dismissal was racially motivated, and

> … pointed out that if Management allowed these sort of pleas to interfere with their decisions, it would be extremely difficult, if not impossible, to maintain any sort of discipline in the Plant and that Management was virtually handing over its right to manage to Government.[39]

Despite the establishment of a Commission of Inquiry into the case, impatience grew about the lack of progress.[40] In early March,

> [ZMU branch] Executive members patrolled Masala location using portable loud hailer … [and] stated "… We thought Government would do something good about Musonda but they have done nothing. We are not afraid to go to gaol. We are not against Government but if we, the Union, ask for something we expect to get it."[41]

Following the failure of tripartite talks, the union defied Ministry pressure and approved the strike.[42] President Kaunda publicly warned:

> "We will not allow the industry that gives us money for development to be paralysed" he said. He hit at the Mineworkers' Union … for going ahead with strike plans after talks with the Government and the setting up of a Commission of Inquiry, "Is that Zambian?" he asked. "I do not accept this. We do not forbid strikes. It is the right of every worker to strike, but they must have some basis … We demand discipline in the Army, the school, the party, industry." [43]

ZMU President Abel Musonda (no relation) responded, 'We did all in our power to avert the strike. We offered quite a few compromises, but the employers would not accept. We have done nothing illegal ... We are just protecting our rights as workers.'[44] The all-out strike at NCR lasted for five days.[45] Threats were made to widen the action to other mines if Musonda was not reinstated.[46] At a ZMU Supreme Council meeting, Labour Minister Mundia

> ... cajoled, begged and entreated the Supreme Council to instruct ... members ... to return to work. He ... appealed for restraint due to the severe financial implications as far as development plans are concerned. The Minister's pleas fell on deaf ears and he received what can only be described as a hostile reception.[47]

After strikers were persuaded to return to work, the union and companies expressed their differing expectations of post-colonial industrial relations at a subsequent Commission of Inquiry. The ZMU criticised the Recognition Agreement, which prevented the Union from challenging dismissals and suspensions.[48] NCR Branch Chairman Marko Lupiya argued that local union officials should be given unfettered access to all areas of the plant.[49] In contrast, the company representative, supporting Minister Mundia's proposal that union officials who do not follow their own Constitution should be dismissed, 'suggested that if the leaders could not stop strikers then they were not leaders.'[50] RST implemented the Commission's recommendation to offer Musonda a different mining job, but continued to resist the freedom of movement of ZMU officials.[51]

Pay negotiations

Under membership pressure, AMWU sought a transformational post-Independence wage increase. The central demand for equal pay was expressed in negotiations in 1965. The mine companies were privately concerned that equal pay for skilled Zambian and expatriate workers would 'pull up' the wages of the general African labour force. Arguing that black Zambian wages were amongst the highest in sub-Saharan Africa, they

> ... explained that expatriate and local rates had to be based on quite different principles. Expatriates had to be sought in the world

> market and offered an inducement to leave their home countries and accept temporary employment in a foreign land.[52]

The union's arguments were indicative of their distinctive vision of a future Zambia:

> African Workers … have contributed greatly to the country's economy. The unfounded fears by the Mining Companies that paying Africans 75 to 80% of the expatriate earnings would disrupt the economy of the country must be discarded … This will … [lead to an] … [i]ncrease in the African purchasing power to the advantage of the industries which will create local markets for perishable products. At present the African only contributes to the growth of the industry by man power and very little by economic purchasing power.[53]

The persistent gap in perceptions of a 'fair wage' provided a significant basis of conflict between the mineworkers, the mine companies and the government. As former mineworker Sam Kangwa puts it:

> … copper was the mainstay of the Zambian economy, and in those days … Zambia was depending on copper for about 98%. So miners … said, we are the people who are producing, we are the people running the country, we are the people who are actually causing the economy of Zambia to tick, so we should be rewarded.[54]

The government pressurised the companies to increase their offer, whilst lobbying for conciliation and seeking to avoid a strike. The mine companies suggested to Minister Mundia that the ZMU was taking advantage of his intervention and was being unreasonable in its demands.[55] The ZMU hoped that UNIP would legislate or pressure the mine companies to introduce a single industry wage structure.[56] In May, a Conciliation Board was established, to which both sides claimed that government policy supported their position.[57] ZMU leaders, however, became increasingly disillusioned with the lack of official support and declared a dispute in July. Mufulira Branch Chairman Fidelis Mwamba complained:

> these people in the house of Assembly have forgotten all the promises that they made before they were put in power, I have got

> the man[i]festo and…we shall go and tell them (MEMBERS) that these promises will never be adhered to.[58]

This discontent was increased by the removal of ZMU President Musonda under the legislation requiring all union officers to have worked in the industry they represented.[59] Musonda, like his predecessor John Chisata, was a UNIP supporter who nevertheless sought to maintain ZMU autonomy. He believes his removal was a politically inspired attempt to undermine this.[60]

Following consultative meetings in the mine townships, the ZMU requested a strike ballot.[61] RTU Axson Chalikulima now informed the ZMU that new branch and national elections were necessary before any strike ballot could take place. Whilst there is no direct evidence that this was designed to prevent a strike, it certainly had that effect. Chalikulima reported on his meeting with ZMU leaders in October:

> The impression was gathered that the Union's Executive was not particularly anxious to call a strike (which would not be favoured by Government), and would prefer to continue negotiations … the Union has not given up the idea of possible strike action … any chance of early action has been put off.[62]

Impatience at this delay was expressed at branch meetings. Company intelligence from Chibuluma reported:

> The crowd as usual was in favour of a strike. There was a Labour Officer present and much criticism was levelled at Government through him for interference in Union matters. When the Union Leaders started to talk about the arrangements for the elections the crowd drifted away shouting that they were only interested in money.[63]

The ZMU subsequently withdrew its strike ballot request in December 1965, in recognition of the difficult political and economic situation resulting from Rhodesian UDI.

ZMU elections, 1965 - 1966

UNIP sought to use the union elections to place loyalist party members in local and national ZMU positions. Bates cites 'warning' letters sent by local UNIP officials to existing ZMU officials, threatening retribution if they stood for re-election against Party-endorsed candidates.[64] He does not, however, adequately describe the extent of intimidation that took place. At an emergency Supreme Council meeting in December, Konkola officials reported that the UNIP Constituency Secretary had told them that any ZMU official who stood would be expelled from UNIP forever and that their UNIP replacements would carry out Government policy.[65] Branch officials in Bancroft, Ndola and Mufulira went into hiding after their counterparts in Butondo were physically attacked by UNIP youths, despite Kaunda's assurances that interference would not occur.[66]

Such threats did not deter branch officials from standing in the elections, where further conflict took place. An election meeting in Chibuluma, where UNIP propaganda claimed the ZMU branch was being run by Malawians, ended in disorder. '*Kasapato*' leaflets made lurid allegations:

> Do you know the present Union Leadership is their [sic] for its own ends … Did you know the Union Head office and the Mining company are working to over throw the Government? … They use your public car to cool down their sexual temperature did you buy it for that? … They plot to kill His Excellency the President, with a Rhodesian gang – are they true leaders of you the people?[67]

Some union officials closed meetings before UNIP could propose candidates, whilst elsewhere, UNIP proposed candidates who were not ZMU members. In Mufulira, the Labour Officer reported:

> As the [ZMU] Regional Secretary spoke, the U.N.I.P. car with loud speakers came and stopped behind the pulpit and the loud speakers blarred [sic] out remarks that the present Union officials were completely ineffective and must be replaced with new blood … The Union Regional Secretary … explained that the Union has for a very long time now been deterred in its efforts to better the conditions of its members by Government … there were angry shouts from the public (union members) that they were not going to tolerate the interference of the party in the Union's activities … The meeting had to be closed on the instructions of the Police as violence was at the fringe of erupting.[68]

After meeting ZMU leaders, Assistant Labour Commissioner Sabel, the senior labour official on the Copperbelt, expressed

> … concern … at the distinct possibility of a general deterioration in the relationships between the Zambia Mineworkers' Union and both Government and the Party throughout the Copperbelt with a real risk of serious outbreaks of violence a[n]d thuggery … ZMU Head Office officials … stressed their great concern at … the hostility of the party at local level towards their organisation and their reluctance on this account to accept assurances at their face value regarding Government's wish to see a strong trade union movement which was not subject to absolute control by Government … they were not merely anxious to co-operate actively with the Government over wages policy but were acutely aware in the present national circumstances of the need to take full account of recent burdens on the economy into their thinking on questions of negotiations … The danger I foresee from any attempt to "pack" the delegate conference from the party side is that it will inevitably consolidate the Union hierarchy who would rather hang together than separately … I think it is feasible to coax the Union further along the road of active co-operation with the Government without the prior need for a drastic surgical operation on its internal composition … the acid test on the Union side will be to see what their concrete policy will be in the immediate future in regard to the current dispute in the mining industry.[69]

Sabel's warning appears to have been heeded. Branch elections were rerun with less disruption in February 1966 and, in most areas, ZMU's nominated officials were elected. At the subsequent delegate conference, the ZMU leadership was returned. Some local UNIP officials were unhappy, but Government leaders accepted the results.[70] Seeking to explain the election, Bates argues:

> In their campaigns against the union, the party leaders emphasized their dedication to the fulfilment, not the curtailment, of the workers' demands. On the part of the workers, the rejection of UNIP's campaign was therefore not an act of resistance to the government or its policies. Rather, it was a selection among competing claimants to expertise as representatives of their interests as mine employees. So long as the union leaders were UNIP, the

> workers were satisfied as to their political loyalty. And once satisfied that all claimants were loyal, they selected those who they felt were most knowledgeable in industrial matters.[71]

This overstates the primacy of UNIP affiliation. Possession of a UNIP membership card was no guarantee of practical political loyalty. Miners supported able leaders who had demonstrated their autonomy from UNIP. In an intimidatory atmosphere, they were not persuaded to vote for UNIP nominees by claims of ZMU officials' alleged disloyalty. The capacity to deliver on industrial demands was indeed the key criteria for electing branch officials, but this was associated with autonomy from Party and Government.

The 1966 pay strike

During the elections, the ZMU suddenly accepted the mining companies' pay offer; basic monthly pay for underground workers rose to £23 7s 6d per month. The ZMU publicly declared that it would make no further pay claim for two years.[72] UDI influenced union leaders' decision to settle, but they had apparently passed Sabel's 'acid test'.[73] Mineworkers, however, were unhappy with the agreement. A strike began in Nchanga on 25 March and spread to all Copperbelt mines by 6 April. Organised by rank and file leaders, the strike was marked by heavy picketing. In Nchanga, 'There was general picketing at the Under-pass to the Plant area on all three shifts…One local employee has been arrested for picketing and detained by the Zambia Police.'[74] Mineworkers' hostility to the agreement was expressed in the manhandling of branch officials.

Under-Minister of Labour Andrew Mutemba announced that the Government had appointed a Commission of Inquiry into their grievances, to be headed by Roland Brown.[75] Despite Mutemba's appeal for a return to work, the strike continued to spread. Kaunda expressed the Government's sympathy to miners' demands in a nationwide broadcast:

> As you are all aware, the mines are the economic lifeblood of the nation and the wealth produced by the mining industry is vital in the struggle against ignorance, poverty and disease. For those reasons I now ask you all to go back to work immediately.[76]

In Chibuluma, a company telegram reported:

> ZMU Leaders organised a Public Meeting this morning and were to have relayed the Presidents speech over Loudspeakers ... One Union Official spoke urging the crowd to return to work but the Platform was taken over by an Employee who advocated staying on Strike stop he said Commissions take months to reach a conclusion which should be obvious to all stop the crowd supported the Rebel Speaker and the Meeting broke up in disorder when Union Officials tried to remove him from the platform. The Presidents speech was not relayed stop[77]

UNIP criticism of the pay deal did not lead miners to respond to its appeals to return to work. At Chamboli, 1,500 miners heckled UNIP Regional Secretary Evaristo Mutale, shouting:

> "We will not go back until we get more money. We would like the President to come here himself." When Mr Mutale said "You have been misled by your union leaders" a section of the crowd replied "You have let us down too."[78]

After further appeals, miners returned to work. The ZMU leadership, previously praised for settling the dispute, was now publicly vilified for accepting such a poor deal. There were renewed calls for increased Government and ZCTU control over unions; Tanganyika, where unions had been brought under full Government control, was cited approvingly.[79]

Branch 'public meetings' were blamed for reducing the union's ability to control its members. The open nature of such meetings, as in the colonial era, provided the most significant forum for the expression of rank-and-file consciousness.[80] Andrew Leslie, leader of the European Mine Workers' Society (MWS), blamed them for poor communications in the ZMU:

> After each agreement reached between the Companies and the Union Representatives these latter are faced with an impossible task of explaining changes in conditions of employment to the massive meetings of thousands at a time of whom two thirds may be members of the Union, the other third being the unemployed, women and political agitators. Since of necessity such meetings are held in the open there is no way of preventing non-members from attending. Under such conditions Union representatives have no

> hope of successfully explaining any situation or putting across any unpopular matter...[81]

The importance of public meetings is further explored in Chapter Four.

The Zambia National Provident Fund strike

By June 1966, criticisms were circulating about the new statutory National Provident Fund (ZNPF), which, unlike the existing 'ZAMINLO' mine pension scheme, did not pay benefits until the age of 55.[82] Mineworkers demanded the refund of ZAMINLO contributions before the ZNPF was introduced. To support their demands, miners in Mufulira walked out in late August. Government ministers accused Europeans and the ANC of being behind the strike, and threatened to detain AMWU officials as their colonial predecessors had in 1956. [83]

All mines were nevertheless out on strike by 2 September. The press renewed its attacks on ZMU leaders for failing to discipline their members.[84] Minister of Labour Munukayumbwa Sipalo told strikers, 'Your strike action is a stab in the back. You have demonstrated your support for Ian Smith.'[85] In Ndola, UNIP Youth Brigades toured Masala township in a government vehicle, ordering miners to return to work.[86] There were reports of physical violence against those trying to do so, including attacks on their properties at night. Police intelligence officers contacted the mines for daily updates:

> Mr. Fleming of the C.I.D. Section called on the Acting Senior Personnel Officer and asked if he could furnish him with names and particulars of suspected intimidators … as His Excellency the President would like to be well informed about the strike situation.[87]

In mid-September, the Government conceded the repayment of ZAMINLO funds.[88] However, many miners initially refused to return until the funds were actually paid. On 17 September, 33 branch leaders were arrested and 'restricted' to rural areas. The return to work gathered pace, but many miners were angered by the restrictions and initially refused to go back. Jonathan Mumba of Luanshya wrote to the *Times*:

> What miners, and all citizens, do not like is to be threatened on matters known to everybody to be reasonable and down to earth. Why should our own ministers say they would jail miners on strike? How many prisons is the Government going to build for such a bulk

> of people? … These things put miners in a very awkward position – they helped in attaining independence and now all of them could be jailed.[89]

A full return to work was secured by 24 September, but the detentions left a legacy of bitterness. Fidelis Mwamba was restricted to his home area in Northern Province.[90] He claims that miners were annoyed by his detention, believing him to be a good leader who had sought to protect their interests.[91] The detainees were allowed to return after six months. Mwamba was transferred to Luanshya where, in a demonstration of his continued popularity, he was elected to the Branch Executive in 1967.[92]

The Brown Commission

Discontent was tempered by the Brown Commission. The mine companies cited the Government's policy of wage control to justify their position. Dr. Petter Bjerve, an economist who had been part of the Seers Report team, told the Commission that unions

> should try to explain to their members the policy being pursued by the Government: that for the sake of Zambia too rapid wage rises would have adverse consequences on the economy. Their role should perhaps be more educational than political.[93]

ZMU President Kossam Milambo told the Inquiry that the strike was justified, admitting he had signed the pay agreement knowing it to be unsatisfactory because of the 'circumstances'. Asked whether he believed in union discipline, he replied, 'It is, of course our duty as trade union officials to discipline our members, but it is difficult if the leaders try to impose on their members something the members do not want.'[94]

The Commission recommended a 22% wage increase for all African miners as a step towards the achievement of African wages equivalent to two-thirds of expatriate wages. Brown also recommended that a single wage scale be (re)introduced for all miners and that 'sub-standard' housing lacking running water be rapidly replaced. The increase was quickly implemented and was subsequently matched by similar increases for other workers. The Brown Commission increases have been portrayed as a generous award, based on UNIP's good relations with urban workers. Bates argued that the Government's pursuit of social justice led it to oppose racially-based wages.[95] Liatto cited the award as evidence that 'government policy was guided first of all by the need to establish a reasonable wage with reference to the living wage and the expectations of workers'.[96]

In fact, as the above evidence demonstrates, UNIP pressure was influential in the ZMU leadership's agreement to a low pay increase in January. The Brown Commission was established only in response to unofficial strike action. This experience, and that of the subsequent ZAMINLO strike, is key to understanding the lessons drawn by miners from these events. For them, the substantial improvements in wages and conditions they expected as the fruits of Independence could be achieved; this depended however not on Government generosity or on negotiations, but (as in the colonial era) on their local organisation and their willingness to take action in pursuit of these objectives, despite opposition from the authorities. As miner Timothy Walamba puts it:

> ... 22% increment ... it didn't come on a silver plate. It came as a result of a general strike ... if you don't protest, if you don't go on strike ... you don't expect to get an increment at all ... we paralysed the mining industry, we went on strike. And that is how the ... management, they decided to agree on the proposals which were made ... It didn't come on a silver plate, it didn't come from the Government at all.[97]

Nevertheless, mineworkers continued to vote for UNIP. The party won every Copperbelt seat in the 1968 Parliamentary elections and Kaunda secured an impressive vote as President. Some respondents report that the vote was less than free, but this is only a partial explanation.[98] Bates suggests that miners separated political and industrial demands. Walamba argues that miners were grateful to UNIP for the African advancement that was taking place:

> ... when I was ...becoming the first Zambian Winding Engine Driver, 1966, miners then started to see ... Zambians being moved into positions of authority. And, you know there was excitement. So people came to realise that ... when pressure is put on, then more Zambians will go into these positions...[99]

Fidelis Mwamba, detained in 1966, not only voted for UNIP, but also sought (unsuccessfully) to stand for Parliament as a UNIP candidate. Because there seemed little prospect for a credible alternative party, Mwamba sought to work within the ruling party to achieve policies that reflected the miners' demands.[100] Electoral support for UNIP amongst mineworkers was contingent on post-colonial advancement and the belief that this would continue if (as Walamba puts it) 'pressure is put on'.

The establishment of MUZ, 1967

The 1966 strikes increased Government's urgency to achieve effective control over mineworkers. The Trade Unions and Trade Disputes Act had introduced a policy of 'one union, one industry'. Against the wishes of the mine companies, there was now increased pressure for a single mine union. The RTU hoped that skilled workers now organised into the Mines Local Staff Association would, with merger, exercise a moderating influence over militant 'uneducated' ZMU leaders.[101] Government and ZCTU pressure ensured that merger went through against significant ZMU opposition in April - May 1967.[102] In a process led by the Ministry of Labour, merged branch committees were created through a partially self-selecting process by local officials. These committees elected delegates to the first conference of the merged Mineworkers' Union of Zambia (MUZ), which in May 1967 chose a new national leadership dominated by former ZMU officials.[103] MUZ's President, David Mwila, was a 31-year old Luanshya miner.[104] He was endorsed by UNIP in the 1966 ZMU elections, but was also popular amongst mineworkers.[105] The MUZ Supreme Council resolved 'That the Union shall co-operate with the Government, the Ruling Party (UNIP) and Zambia Congress of Trade Unions'.[106] The new General Secretary Edwin Thawe stressed that

> STRIKE ACTION ... should be used only at the last stage...we should educated [sic] our members to obey the leaders and follow the orders. It is now almost a tendancy [sic] in people that they are not prepared to listen to the leaders ... illegal action cannot be supported at all...[107]

Mwila argued that mineworkers' interests and those of Zambia were the same:

> To day [sic] in the new world it is a duty for every true Trade Union leader to comply with the national development ... higher productivity is not an end in itself but a means of Social Progress ... It is therefore in the interest of a working class, particular[l]y of a developing country like Zambia that the national wealth should be boosted so that they can have a larger share of the national cake.[108]

This did not, however, prevent MUZ making new demands. In 1967-68, MUZ sought to ensure that all Brown's recommendations were implemented, particularly an end to discrimination in service provision. Whilst the companies defended these inequalities on the basis that skilled Africans and expatriates expected preferential conditions, MUZ argued that 'Any system dependant on class distinctions was unacceptable to the Union.'[109] Agreement was reached in April 1968, with MUZ winning some of its demands. Achieved without industrial action, the agreement was celebrated as an example of responsible industrial relations.[110] Negotiations took place in the Mines Joint Industrial Committee (MJIC). From this time, however, the MJIC Permanent sub-committee (PC), established in 1964, emerged as the main venue for practical negotiations, with reference back to the full MJIC for formal approval. This represented significant centralisation: whilst all MUZ branches were represented on the MJIC, only three senior Head Office officials attended the sub-committee. Minutes of these meetings indicate an increasingly cooperative relationship between management and the union leadership.

Industrial action, 1967 - 1969

In the wider economy, Government was increasingly concerned about the persistence of strikes. Hundreds of local work stoppages lasting hours or days took place, demanding wage increases and challenging racist supervision. These were generally organised by workers themselves, with national union leaders ordering their return to work. Kapferer found that Kabwe textile workers resisted union intervention in disputes, believing that union leaders and government labour officers conspired in negotiations which did not address their grievances.[111] Major stoppages occurred on the railways (1965, 1968 and 1970), in local councils (1968) and in schools (1970).[112] Politicians periodically proposed compulsory arbitration or a ban on strikes.

In April 1967, at a major tripartite conference on labour, Minister of Labour Sipalo attacked workers' laziness:

> You and I know that if you were to put workers on the road to work on their own without a whip behind them, you would find them seated in the next ten minutes ... I can assure you that this is the responsibility of trade union leaders just as it is the responsibility of civil service leaders and political leaders.[113]

The Conference agreed to link future wage increases to productivity. Union leaders pledged to put Zambia's interests before their own and to 'use all our efforts to maintain industrial peace.'[114] The level of strikes did not, however, diminish.

The emphasis on productivity was reinforced in an influential 1969 ILO report on wages and incomes.[115] H A Turner's report linked rising wages with rural stagnation, producing data (later challenged by Fry) to demonstrate a fall in labour productivity since Independence.[116] Turner argued that 'the colonial system of labour discipline has broken down and nothing has yet developed to take its place.'[117]

Union leaders, however, rejected Government claims that they indirectly supported strikes, arguing that branch leaders organised them to make themselves popular.[118] Sociologist Alexander Matejko agreed:

> the lower echelons of the trade union hierarchy, not being professionally settled and relying to a large extent on their popularity among the rank-and-file, might even have a vested interest in instigating unrest and unofficial strikes. It gives them a certain sense of power, and sometimes leads to their promotion to headquarters.[119]

Whilst advocating Government intervention, Matjeko admitted that 'Trade unionism constantly patronised by the Government and closely controlled by it will lose both its autonomous character and at the same time its appeal among the rank-and-file.'[120]

African advancement and Zambianisation

In the late 1960s, conflicts in the mining industry centred on contesting definitions of African advancement, often replicating those identified by Morrison in 1962 (see Chapter One). Aaron Milner, Minister in the President's Office, chaired the national tripartite Zambianisation committee, overseeing the replacement of skilled expatriates with Zambians. Milner utilised UNIP's characteristic approach of public exhortation and private lobbying, whilst rejecting direct interference. The ZMU cited Brown's recommendations in demanding that 'ultimate responsibility for reaching decisions about the replacement of expatriate workers by Zambians should rest with the Committee and not with the Management of the Mines.'[121] Milner however argued that

> ... interference by the Committee in the management of a mine could lead to loss of production ... and assured the Mining

> Companies representatives that no decision made by the Committee would be to [t]he detriment of the people or economy of the country.[122]

The mine companies, whilst training Zambians for skilled posts, rejected rapid Zambianisation on the basis that it would impair production and thereby reduce income to the state; this, for the Government, remained the overriding concern.

For mineworkers, 'advancement' also meant progress towards equal pay for the same job. For groups of newly promoted workers, such as Winding Engine Drivers (WEDs), Zambianisation was a matter of both pride and anger, as they were paid one-third of the salary of those they replaced. Such groups, seeking salaries reflective of their new position, pursued them through localised strikes. Advancement was also an arena of conflict for the majority of unskilled workers in an increasingly capital intensive industry, where the previous rapid expansion in the labour force was curtailed, severely limiting opportunities for promotion. As MUZ argued:

> … unrest was largely caused by the younger better educated ambitious men who wanted to make mining a career. The only way they could get into the industry was by accepting work as Lashers [unskilled underground workers], but they expected rapid promotion. When their expectations were frustrated, the[y] became discontented, and their discontent spread to other workers.[123]

Burawoy described the impact of the breakdown of coercive labour supervision:

> … section and shift bosses no longer had the support of colonial sanctions and apparatuses. To be sure they had the power to allocate work and enforce overtime, but lashers were able to resist by manipulating the work situation, pretending to be sick, and so forth. A situation of continuous struggle ensued, leading to spontaneous walkouts and sometimes to wildcat strikes.[124]

Striking lashers demanded both rapid promotion and higher wages for their current grades. Unskilled mineworkers were angered by the new grade-based discrimination in the provision of housing and other services.

This, as Burawoy identified, created an increased sense of class awareness.[125]

Local strikes, occurring two or three times a month across the Copperbelt, were combated by the increasing arrest of pickets. MUZ General Secretary Thawe was in constant communication with company officials, discussing the causes of strikes and their resolution. Branch leaders were caught between pressure from the national union and mine management to secure a return to work and the refusal of striking workers to do so. They commonly sought to persuade strikers to return to work so that negotiations could begin, in which they would sympathetically represent their members' grievances. At no stage did mineworkers demonstrate notable willingness to pursue their grievances through the established negotiation process. All strikes were both 'unconstitutional' according to MUZ rules and illegal.

MUZ members increasingly criticised their leaders' moderation. During a strike in Nchanga in 1968, Mwila joined Government in appealing for a return to work:

> Hecklers in the crowd muttered allegations that their union leaders were in connivance with the management … when Mr Bulawayo [Assistant Minister of State for Western Province] attempted to introduce Mr Mwila, the men shouted him down saying they did not want to hear him.[126]

The job evaluation agreement, 1968 - 1970

MUZ tabled a new wage demand in 1968, seeking to bring African mineworkers closer to Brown's 'two-thirds' target. The companies again rejected these claims by reference to Government policy, arguing that 'Government Ministers had stated … there should be wage restraint during the Four Year National Development Plan … It was the duty of employers to be guided by Government policy...'[127] MUZ denied that this was an accurate reflection of Government policy, but sought to avoid industrial action. Pressure for action nevertheless built up at a local level. In March 1969 MUZ declared a dispute.

Tension was increased by the 1969 Budget; this sought to alleviate economic stagnation by taxing many workers for the first time, thus increasing demands for a compensatory wage increase. Two thousand miners stormed out of the meeting in Kitwe when the national anthem was played, declaring they would close the mines 'so the Government feels the pinch.' Local threats were made of strike action to begin on 1 April, when the new tax rates would come into force.[128] As this date approached, the Government appointed a conciliation board. Vice President Kapwepwe appealed to miners not to strike, explaining that the tax reforms were designed only to affect higher paid workers.[129] Kapwepwe, already popular, won renewed respect by fearlessly arguing his case to meetings in the mine townships, reflecting values that miners respected in their leaders. No strike took place. Nevertheless, at conciliation talks MUZ claimed it was

> under heavy pressure from their members to secure an early settlement, and … expressed the view that a contented labour force is the best guarantee of prosperity, and that a withdrawal of labour would do more harm than good to the industry and to the country than the granting of their demands.[130]

Conciliation led to a new job evaluation process, meant to ensure a scientific and neutral assessment of every mine job, addressing disputed grades such as the WEDs. In return, MUZ agreed to work to improve output, reduce indiscipline and absenteeism and punish members initiating wildcat strikes.[131] Job evaluation took far longer than expected. Union members were praised for their patience, but groups of workers struck to demand upgrading.[132] In March 1970, Rokana WEDs

> walked to the headquarters of the Mineworkers' Union of Zambia today and told MUZ President that unless they received a reply to their complaint they would not report for work. The drivers paraded a poster reading: "Expatriates receive K400 per month for doing the same job that we are doing. But we are receiving K147. The gap must be bridged."[133]

Anxiety regarding delays was expressed at a MJIC PC meeting by Minister of Labour Fwanyanga Mulikita:

> The Minister said … There had been rumours that mineworkers were likely to stage a demonstration strike in protest at the delay in commencing negotiations. There had been allegations in the press that the workers … had even accused their own Union leaders of employing delaying tactics. The Minister said he wished to assure the Companies and the Union of his full support in measures to forestall irresponsible action which might jeopardise the whole exercise…Mr.

> Thawe said that the Union executive had been successful in restraining its members so far, but he was afraid that the executive could not continue to do so if there were further delays.[134]

An agreement was signed in July 1970.[135] In it, MUZ agreed to new measures to combat absenteeism, particularly the withholding of sick pay for the first two days. A few thousands of miners received a wage increase from regrading, but most, including the WEDs, did not. WEDs now threatened to leave MUZ.[136] One wrote:

> Immediately a Zambian took over this job, his salary was slashed to less than half the salary of an expatriate driver. This was not even worked out from the so-called local conditions, but was merely an individual's practice of economics and accounts with those yes-men from the corrupt MUZ. How can you call this Zambianisation when after taking over the job from an expatriate, you don't qualify to live in the house he had? [137]

A subsequent agreement awarded a 5% annual wage increase for 1969 and 1970 to all miners.[138] This was not, however, sufficient to address discontent; indeed, it merely presaged a substantial challenge to the MUZ leadership.

Nationalisation and the Industrial Relations Act

Changes to industrial management and ownership significantly increased the role of the Government in labour relations. In 1968, 26 companies were nationalised, with compensation. In August 1969, in his speech announcing the 51% nationalisation of the copper mining industry, Kaunda claimed it would create a 'classless society' that would allow ordinary Zambians to control their economy.[139] This had important consequences for industrial relations:

> The State … holds industrial investments, not for its own good, <u>not</u> merely for the good of those directly employed in the State enterprises, but for the benefit of Zambians everywhere. Thus, for a union to push a claim against the State is to push a claim against the people…[140]

In this new system, workers could no longer pursue narrow economic interests at the (supposed) expense of their rural cousins. Kaunda condemned wildcat strikes and, (in the same speech) introduced an immediate wage freeze and a temporary ban on all strikes.[141] Bates noted approvingly that mine nationalisation 'clarified the link between the conduct of labor and the success of development efforts, and strengthened the government's resolve to curtail the militancy of labor...'[142] Burawoy argued that 'the proposed nationalisation … cemented [the companies'] co-operation and identification with the Zambian government, giving them much greater security if faced with opposition from … their black labour force.'[143] Certainly, one mine company had first welcomed the idea of Government as a shareholder in 1964.[144] As *The Economist* put it:

> … the shrewdest businessmen in that part of the world have argued for some time that 49 per cent stake in a business whose success is underwritten by government participation may be more valuable than 100 per cent of a concern exposed to all the political winds that blow…[145]

The issue at hand was not ownership but control of the new mining companies, RCM and NCCM. Despite Kaunda's claims regarding Zambian control, AAC and RST received a 'Categorical statement that takeover of mines was between Government and shareholders only, and would not affect employees, management and running of mines at all.'[146] Nationalisation aimed to increase both the effective control of strategic natural resources and the human resources vital to their utilisation.

At the same time, UNIP sought new methods to control industrial unrest, whilst apparently offering worker participation in management. Draft legislation proposed new Works Councils with legal rights of veto and consultation over some management decisions. Government and ZCTU leaders agreed that

> … participation in … management would be fulfilling the aims of Independence. Economically, there would be equal distribution of wealth and, socially, participants themselves would feel proud because they would be aware that they were contributing to the economic development of their country.[147]

Kaunda however criticised the draft legislation for limiting the powers of the ZCTU:

> It would appear that the Draft was written from the point of view of what the Labour Commissioner will require in terms of power and authority in order to control the trade unions as effectively as possible. My proposals, on the contrary, were clearly intended to introduce a system whereby the trade unions would be enabled to administer their own affairs and discipline and control their members...[148]

Thus, a public debate marked by claims of worker empowerment is revealed as a tactical debate regarding the best way to control workers. Given the loyalty of the ZCTU, it is unsurprising that Kaunda believed that a Congress with enhanced powers over its members could and would play such a role. Kaunda ensured that the ZCTU's powers over member unions were substantially increased in the resultant 1971 Industrial Relations Act. The Act also introduced a system of conciliation and arbitration that made it effectively impossible to call a legal strike, a decision criticised by union leaders.[149]

Lobbying by business representatives had, meanwhile, successfully limited the effective powers of Works Councils.[150] Minister of Labour Mulikita privately assured the mine companies:

> ... the approval of the Works Councils will be required only as regards decisions by management on matters of policy in the field of personnel management and industrial relations and on no other matters.[151]

The mining companies approvingly assessed the resultant legislation:

> The Works Council operates at the highest level in the undertaking. The rank and file will not feel any sense of participation in management ... the professed socialist-humanistic philosophy has not influenced the attitudes of workers and management nor had it altered the capitalist structure of the country's productive economy.[152]

The 1970 MUZ Conference

During the job evaluation process, public branch meetings heard increasing complaints that Head Office leaders were in connivance with management.[153] The ban on strikes announced by Kaunda in August 1969 was briefly effective, but Rokana lashers struck in January 1970, expressing no confidence in the MUZ leadership.[154] Union elections due in May 1970 were postponed to allow the job evaluation agreement to be completed. The agreement was subsequently attacked by Luanshya branch leaders, including Kossam Milambo and Abel Musonda, who also criticised the MUZ leadership's closeness to Government and the mine companies. MUZ President Mwila's appointment to the board of the nationalised mine company in January 1970 was celebrated by MUZ as an example of workers' participation in industry.[155] However, critics like Luanshya rank-and-file leader Campbell Chibamba argued that 'it is impossible for him to function effectively as a Mindeco director and at the same time be a true representative of employees.'[156] In response, MUZ's Supreme Council removed Milambo from his position as MUZ Vice President and expelled four Luanshya branch leaders, calling them 'traitors to industrial stability.'[157]

Pressure increased to hold the delayed MUZ elections.[158] Branch elections finally took place in October 1970. Abel Musonda, having declared he would challenge Mwila for the MUZ Presidency, was elected as Roan Branch Chairman, with his supporters winning other positions.[159] General Secretary Thawe accused the Luanshya leaders of trying to turn the union into a political party.[160] Days later, Musonda was 'offered' a position as District Governor, which he initially rejected.[161] In November, national elections took place at the MUZ Conference, in which MUZ's own employees were controversially allowed to vote by the Labour Commissioner. Musonda led a walkout of ten of the 62 delegates in protest and Mwila was re-elected unopposed.[162] There were now rumours that Musonda, still ZCTU Vice President, would lead a splinter mineworkers' union. MUZ leaders called for Musonda's removal from his ZCTU position and the windows of his house were smashed. In mid December, Musonda agreed to take up an offer he says he could not refuse:

> Kenneth Kaunda ... said look, I am the President of the Republic. I've got the power to appoint. So, whether I like it or not, you are going to take up this appointment ... I said no I'm not going to leave...Until, he directed the Management at Roan to ... fire me. And in the end I was going to become a destitute ... Before accepting, I approached ... Simon Kapwepwe. I said "Look, old man, the President has appointed me ... District Governor. I am not willing to ... take up that appointment." He said that, "Young man, please take up that appointment. For your own good. Or

> refuse to take up, at your own risk." … I had no choice but to take up the appointment as District Governor.[163]

Dissent in MUZ, 1970 - 1971

A new dissident leadership now emerged. In March 1971, Palakasa Chiwaya was removed from his position as Chililabombwe Branch Chairman at a secret meeting. Press coverage suggested that Chiwaya's rank-and-file support was based on his opposition to the job evaluation agreement.[164] At a mass meeting, 1,000 miners declared they 'would fight "tooth and nail" to remove the national leadership.'[165] Later that month, Nkana Branch Vice Chairman Roy Yumbe was suspended. The *Times* reported:

> Already a group calling itself the "Silent Burners" has been formed to challenge his [Yumbe's] eight-month suspension … "Burners" spokesman Mr Jameson Bwalya said there were plans to hold massive demonstrations … The leadership of MUZ head office had become "questionable" when they signed an agreement which made miners lose two days pay if they were admitted to hospital. The group supports controversial dissident MUZ leader in Chililabombwe, Mr Palakasa Chiwaya.[166]

The extent of support for this dissident group is hard to ascertain. Yumbe claims the support of the majority of Rokana miners:

> the Government … was not anxious to give … miners what they wanted … you see, the politicians, they promise, they want to be voted into office … and after being voted they forget what they … have been promising. So, what happened to the miners, they were angered … So … me, and my colleague Supaile … we started … telling the miners … the truth. So we found that the Government connived with the Union, and most of the people, miners, followed us, me and Supaile.[167]

Certainly, newspapers cited the high level of support for the dissidents as evidence that the job evaluation agreement did not have the support of most miners.[168] In April, anti-Mwila posters appeared in Chililabombwe. 500 Nkana miners attended a secret meeting, where a petition was launched against Mwila, demanding fresh elections.[169] MUZ dissidents appear to have initially rejected militant tactics and focussed on the narrow issue of Mwila's leadership, hoping for a sympathetic approach from the Government. This was reciprocated; Deputy Labour Minister Timothy Kankasa held secret talks with Chiwaya and the MUZ head office officials where their differences were resolved.

Opposition nevertheless continued. One thousand miners signed the anti-Mwila petition and Supaile publicly attacked Mwila as Chakulya's puppet.[170] In Chililabombwe, thousands of mineworkers shouted down Mwila and Thawe with cries of 'sell out'.[171] Supaile then declared a boycott of the imminent May Day celebrations. As noted above, May Day was a cornerstone of the public expression of unity between the labour movement and the state. The boycott could not help but place the 'rebel' group in conflict with the Government. Some May Day events were cancelled because of poor attendance, providing apparent evidence of support for the rebels. Kankasa now directly criticised Supaile.[172]

Despite this, the dissidents were still willing to appeal to the state. On 7 May 1971, Supaile wrote to Minister Mulikita, 'for and on behalf of other miners', providing an eloquent expression of mineworkers' discontent:

> Mining is the most hazardous and dangerous operations in Zambia … Zambian miners are amongst the most hard working members of the community. Production on the mines has increased since 1964 and the Zambian has contributed to this increase in production.
>
> Zambianization and the introduction of the local wage structure has served [sic – saved] the Mining Companies millions and millions of Kwacha in salaries and services. A job which was done by an expatriate at a very high salary and housed in an expensive house is now done by a Zambian at less than one third of the expatriate's pay and housed in a hut which the Brown Commission condemned as unfit for human habitation … The Brown Commission also strongly recommended, among other improvements, a single wage structure and similar conditions of employment for all employees in the Mining Industry.
>
> The present agreement … was reached and signed without the slightest consideration to most of the proceeding … factors. The … agreement [was] discussed and agreed upon in private and … [the deduction of two days' sick pay] … angered miners so much that the present Union officials cannot safely talk to them about the

> agreement and affairs of the Union in general. The Union officials are afraid of their members…[173]

Supaile cited ten breaches of the MUZ Constitution to show that the Head Office leadership was elected unconstitutionally:

> To a large extent the Labour Day boycott can be attributed to the Zambian Miners demand that the present union officials should step down … we are hereby formally asking the Minister of Labour and Social Services to take appropriate action and dissolve the present Supreme Council headed by Mr. David Mwila and arrange for fresh elections throughout the entire union.[174]

In May, Luanshya WEDs struck against the job evaluation agreement.[175] Two days later, representatives from all mines met in Chambishi and threatened a new strike on 2 June in an effort to oust Mwila, criticising the agreement for not delivering equal pay for equal work.[176] Government and the companies shared their anxieties about the situation, but each sought to persuade each other to take responsibility for the problem. On 13 May, company officials met Simon Bwalya, the senior Ministry official on the Copperbelt:

> Bwalya … said the purpose of his visit was to enquire how seriously the companies viewed the situation, and whether they intended to take discipllinary [sic] action against dissidents … We said that we took a very serious view of the situation … It would be wrong, however, for the companies to become involved in a dispute within the union. We considered that the union head office leadership had shown considerable responsibility in the manner in which they negotiated the 1970 agreement, and that the attack on those leaders was the first stage in an attempt to overthrow the agreement. We pointed out that if this was allowed to happen, the future of collective bargaining, not only in the mining industry but in the whole country, would be jeopardised. Nevertheless any attempt by the companies to support the union leaders would give the dissidents ground for claiming that the leaders were mere company stooges…
>
> We said that the dissidents would take advantage of every incident, and unless action was taken to stop their activities we thought it

> possible that they could build up sufficient support to overthrow the leaders … the position was very serious … some action was probably required from Government to curb the activities of the dissident group, so as to prevent a situation whereby the mines and therefore the economy of the country was affected.[177]

Three days later, Supaile, Yumbe and 13 other 'dissident' leaders from across the Copperbelt were arrested and restricted to their 'home' villages. One hundred striking WEDs were arrested and released only when they agreed to return to work. Mwila won a vote of confidence at an 'extraordinary conference' of all MUZ branches.[178] The *Times of Zambia* reported:

> UNIP itself was involved in the business on the Copperbelt. Posters urging the rebel miners to strike and halt production at Rokana in Kitwe were torn down by UNIP members in the townships. And UNIP leaders … did not hesitate to warn the miners that if they went on strike they would be answerable not only to the Government and the forces of law and order, but also the party itself.
>
> Mr Mwila's troubles have already drawn the Government into the MUZ orbit in a manner which could earn the Kaunda administration a bad name among the miners, many of whom are staunch UNIP members and hold top posts in the party's branch, constituency or regional offices on the Copperbelt … the Government's decision to remove Mr Abel Musonda from his mine job in Luanshya was calculated deliberately to reduce the intensity of opposition against Mr Mwila.
>
> The more they came up against the new conditions created under the job evaluation agreement, the more they were convinced that they had been sold down the river. Mr Mwila's name became synonymous with the agreement and he was openly accused of betraying his colleagues. He had to go, most of the miners thought.[179]

In the run-up to the strike threatened for 2 June, UNIP youths toured mine townships looking for strike agitators. MUZ leaders called the strike 'futile' and Copperbelt Minister Alex Shapi warned that strikers would be prosecuted. This pressure proved effective: the proposed strike did not take place.[180] With the removal of its leadership, the dissident group was unable to effectively resist the authority of the state and UNIP on the Copperbelt. However, as the *Times* feared, this experience led many miners to question their support for UNIP. This coincided with the biggest political threat to UNIP supremacy since Independence.

The United Progressive Party

Most analysts of the UPP's challenge to UNIP have tended to accept its successful suppression as evidence of its lack of support. Gertzel argued that, 'The miners have always regarded themselves as loyal to UNIP; they have (with the exception of one element in Mufulira) consistently supported UNIP…'[181] Burawoy only notes UPP support amongst junior mine personnel officers.[182] Gertzel, however, perceptively noted that in 1971–72:

> … the loyalty of the labour movement, including the miners, became crucial not only for the government's position but for the maintenance of authority itself … the danger of the opposition [UPP] working within the labour movement and of winning its support was apparently considerable; in that situation they were bound to seek greater control of labour leadership. The extent to which government control of the union leadership will create a responsive labour force clearly depends on the relationship between that leadership and the rank and file.[183]

Bates identified the link between mineworkers' economic discontent and their support for the UPP:

> Working clandestinely and utilizing their numerous ties with the dissident mineworkers and their spokesmen, the leaders of the UPP hoped to take advantage of currents of dissent in the mine-working community and to gain support for their opposition party.[184]

Bates cites a *Zambia Daily Mail* report confirming that such efforts were successful:

> … dissident miners, angered by the government's roundup of rebel leaders earlier this year, had pledged their support for the new party, called the United Progressive Party … there had been secret

> movements among the miners' dissident groups. The sources said that the rebel miners had promised the organizers of the new party that they would back it…[185]

Bates suggests, and my research confirms, that whilst the MUZ Supreme Council unanimously reaffirmed the union's support for UNIP, a substantial part of the mining community was supportive of and responsive to the ideas of the UPP.[186] Following the party's launch in August 1971, the UPP demanded faster Zambianisation and criticised the favouring of whites in mine housing. Bates describes the way in which UNIP structures were fragmented and divided throughout the Copperbelt, as party organisers (many of them mineworkers) defected to the UPP. The extent of violent clashes between the parties, and of state repression and subsequent banning of the UPP indicate that UNIP's concern about the party was rooted in its presence in the economically pivotal Copperbelt. UPP leader John Chisata describes the ways in which he and other UPP leaders were able to play on mineworkers' discontent with the pay agreement to gain support. He claims that other 'parties did not give them [UNIP] as much worries as us. Because … we were capable, of even … organising … the general strike.'[187] Chisata argues that the UPP, aware that the MUZ leadership was under the influence of UNIP, focused instead on gaining the support of the union's membership:

> when you want to win the support, don't go to the leaders … Leaders are very few, and they're easy to blame … But when you go to the members, taking the members away from any leader … it's like you take water away from the fish…[188]

The ruling party dealt ruthlessly with the UPP. Chisata and most other UPP Copperbelt leaders were detained in mid-September.[189] Fidelis Mwamba, a UPP organiser, found it impossible to hold public meetings in Luanshya, organising them secretly in the 'bush' nearby.[190] Basilio Kunda, MUZ Nchanga Branch Treasurer and UPP district publicity secretary, organised meetings at night, for fear of UNIP 'spies'.[191] MUZ Head Office suspended shop stewards identified as UPP supporters, arguing that this breached MUZ's policy of cooperation with UNIP.

In these circumstances, the UPP sought to campaign for its by-election candidates in the Mufulira West and Wusakile-Chamboli mining constituencies. There were widespread allegations that local UNIP officials had campaigned for the UPP. At least 90% of the population of Mufulira

West were mineworkers or members of their families and therefore made up the vast majority of local UNIP officials. UPP, denied permits to hold public meetings, instead campaigned by word of mouth, organising secret house meetings in the early hours. UPP organisers carried Bibles to indicate they were attending religious gatherings.[192]

In contrast, UNIP's strategy reflected its wider approach to the union movement, relying on MUZ leaders to secure the support of mineworkers. UNIP placed senior government ministers in Mufulira for the campaign and selected Alexander Kamalondo, Mufulira MUZ Branch Chairman and national MUZ Vice President, as its candidate. Kapwepwe won the Mufulira West by-election on 20 December 1971, receiving 2,120 votes against Kamalondo's 1,814.[193] UNIP, however, secured Wusakile-Chamboli by 7,637 votes to 1,445, on a significantly reduced turnout.[194]

The role of MUZ Head Office in UNIP's Mufulira campaign is demonstrated in a 'Confidential' report:

> Mineworkers' Union of Zambia was very grateful to [UNIP] for putting forward their Vice President … to stand … in response they nominated two of their leaders … a car, megaphones and public speaking system to be used in the campaigns.[195]

UPP circulars, it was noted, 'strongly criticised the personality of Comrade Kamalondo, his leadership in M.U.Z. in relation to the new Mine – M.U.Z. Agreement … Although UNIP campaigners tried their best to explain the … Agreement … much lies were already planted in the minds of the miners…'[196] The report also blamed local MUZ officials for UNIP's defeat:

> … we as MUZ campaigners and the Party UNIP were very suspicious, as among the officials is Mr. H. Chanda who is "Chief Shop Steward" but a unquestionable UPP supporter, and has since been suspended from MUZ leadership.[197]

The report complained that card-checking campaigns by UNIP youths had been undermined by the police and recommended that 'The youths should be left free to organise … without any Police or administrative threats…'[198] Thus, in a constituency where a majority of its members had supported an opposition party, MUZ Head Office complained that insufficient pressure had been applied to these members to ensure their loyalty to UNIP and recommended that this be more firmly applied. In a constituency that AMWU President Chisata had won in 1964 because he was associated with the aspirations of mineworkers for Independence, the defeat of MUZ's Vice President resulted from his association with the failure of those aspirations to be met.

General Secretary Thawe subsequently admitted that, 'since the formation of UPP there were many people who were suspected and this had involved the Union very badly.'[199] Following the banning of the UPP, a split occurred in Nkana branch 'as a result of the formation of the banned party.'[200] Subsequent expulsions re-established the leadership's control of the union.

Opposition to UNIP amongst mineworkers nevertheless continued. In 1972, Luanshya branch officials nominated, and mineworkers elected, four candidates for the Roan-Mpatamatu mine township board whom UNIP described as 'former UPP men.'[201] One UNIP official claimed he found local branch leaders

> … together with the so called independent candidates after they had won their election and were celebrating … there were shouts of "UNIP is finished in Luanshya and even if you bring the One Party State Democrancy [sic] we shall still defeat you."[202]

Support for those who had challenged MUZ and UNIP leaders endured at branch level. By June 1972, all those detained in May 1971 except Albert Supaile had been released. Despite Head Office opposition, Yumbe was reinstated as Nkana Branch Vice Chairman.[203]

Towards the one-party state

The appointment of ZCTU leader Wilson Chakulya as Minister of Labour in April 1971 epitomised UNIP's dominance of the official labour movement. In his new position, Chakulya guided the Industrial Relations Act through Parliament. He also appointed accountants to run ZCTU and member union finances, disregarding civil servants' advice that this would be seen as 'gross interference in the internal affairs of the Congress.'[204] Nevertheless, addressing the ZCTU General Council, he rejected 'irresponsible and baseless' accusations that the Government was interfering with the union movement:

> To say that the Z.C.T.U. is affiliated to UNIP because it has not choice implies that the UNIP Government does certain things which are anti-workers, or that if there was another party that party

> might have been more progressive … You, Union leaders must remember at all times that you are Zambians first and foremost and your trade Union positions are secondary … Zambia is a developing country and nobody should be allowed to subvert the trade union movement for political or factional interests.[205]

At least one Council member demurred, arguing that

> … there were areas where the Zambia Congress of Trade Unions and the Party cannot see eye to eye because of the different interests each organisation has … it was wrong to commit the Zambia Congress of Trade Unions into supporting the Party for ever because it was false to say that there can never be any progressive party in future.[206]

However, it quickly became clear that the new ZCTU President, Newstead Zimba, would adopt a more autonomous stance. Zimba (the first ZCTU leader to be freely elected) made it clear that 'Trade unions will not sit idly by, a[n]d watch with connivance workers' rights being eroded.'[207] ZCTU officials criticised the presence of police 'spies' at their meetings and political interference in their affairs.[208] It was becoming clear that the newly strengthened Congress could no longer be taken for granted.

Conclusion

During Zambia's First Republic, declarations of loyalty by union leaders and electoral support for UNIP amongst workers should not be equated to their practical support for the new government's policies, which were based on an accommodation and effective partnership with multinational mine capital. Supportive relations between Government and some union leaders (many appointed by that Government) did not extend to most union members, who sought to ensure their movement's autonomy from political control and criticised leaders seen as too close to the Party and Government.

Mineworkers expected Independence to end the racial inequality and violent supervision that lay at the heart of colonial industrial relations and to extend the practical authority and effectiveness of union representation. In 1966, mineworkers secured significant wage increases in the face of opposition by mine companies and despite state repression of branch leaders. Their demand for equal pay for equal work was, however, not achieved. The mineworkers' union was initially in constant conflict with UNIP. The failure of Government attempts to achieve direct control over the union led to a more indirect approach in the late 1960s, utilising the MUZ and ZCTU's central bureaucracy to restrict membership demands for transformational improvements in wages and conditions. Whilst this task was performed, the resultant anger of mineworkers coalesced into significant rank-and-file opposition to the MUZ leadership, which challenged the union's close relationship to management and its implementation of Government policies. State and party repression of this group addressed its immediate challenge, but did not erase the questions it raised regarding the legitimacy of the post-colonial economic settlement.

Mineworkers' electoral support to UNIP was based on their belief in its promises of post-colonial change. When the limits of this change were demonstrated and when industrial methods of advancement were blocked by legislation and detention, many supported the UPP's populist challenge to UNIP, with some local MUZ leaders detained as a result. Whilst this political challenge was addressed, the repression of dissent delivered only temporary quiescence. The enduring vitality of rank-and-file organisation and the continued willingness of mineworkers and their communities to challenge the practical implementation of UNIP policy were to prove central to the inability of the one-party state to achieve effective control over workers through the agency of the union bureaucracy.

4

TALKING IN DARK CORNERS, 1973 - 1981

Introduction

The establishment of Zambia's one-party state in December 1972 appeared to ensure the political hegemony of UNIP and in particular its control of labour organisation. Baylies and Szeftel's analysis of the one-party state elections of 1973 and 1978 suggests a lack of significant opposition to UNIP on the Copperbelt.[1] Throughout the 1970s, the ZCTU and MUZ consistently proclaimed their public loyalty to the Party and its Government. By the mid 1970s, Gertzel was convinced that 'Recruitment to union office is therefore ultimately controlled by government...'[2] One radical observer concluded that 'the governing class has succeeded in co-opting the labour movement into the state structure in a way which renders its official representatives quite incapable of rending support to the basic demands of the workers they are supposed to represent'.[3]

On the surface, the 1971 Industrial Relations Act (IRA) and the nationalisation of industry seemed to have delivered, if not loyalty, then at least working-class acquiescence. The newly strengthened ZCTU's programme of workers' education was designed to train 'responsible' union branch organisers, who would ensure industrial discipline and productivity. There was a dramatic fall in the number of strikes and days lost in the 1970s, and increased punishments for those that took such illegal action. The MUZ negotiated three successive triennial pay rounds that stayed within legal pay limits and did not lead to industrial action.

Beneath the official surface, however, there was significant grassroots discontent with the new 'industrial participatory democracy'. As this chapter will show, this expressed itself in MUZ branches, and mineworkers' communities, in a range of ways. Challenges to MUZ's leadership, stifled by state intervention (including the use of detention), nevertheless continued to be expressed by branch officials. With the increased repression of overt industrial action, mineworkers turned, as their predecessors had done, to alternative methods of organising strikes.[4] The wives of mineworkers increasingly challenged the authority of the companies and the Government through direct action, influencing the style and content of protests by the mineworking community.

As the copper price stagnated and then declined rapidly from 1975, increasing economic problems led to policy divisions within UNIP. This partly reflected the emergence of a state capitalist class which emerged from the overlap between Government, Party, and parastatal organisations. This class was criticised by the increasingly prominent ZCTU leadership of Newstead Zimba and Frederick Chiluba. Whilst consistently declaring their loyalty to UNIP and Kaunda, they reserved the right to represent and publicly comment on the 'industrial' interests of the labour movement. At a stage removed, these leaders reflected discontent amongst urban workers with the rising cost of living and shortages of essential commodities.

Unable to contain their members' discontent, the MUZ national leadership was increasingly forced to express aspects of it. In 1980, the reorganisation of local government in the mine townships became the focus of direct conflict between the UNIP Government and the labour movement in general, and mineworkers in particular. This culminated in 1981 in a series of strikes, the suspension of union leaders from UNIP and the detention of ZCTU and MUZ leaders. Whilst this conflict has been explained by Mijere as a prime example of the aristocratic nature of Zambia's mineworkers, this chapter argues in contrast that the 'decentralisation' of local government represented an attempt by UNIP to impose its authority over the enduring autonomy of the mine townships. In resisting this, mineworkers and their communities rejected the political and economic assumptions underpinning the one-party system.

The new ZCTU leadership

Throughout the 1970s, the ZCTU consistently declared its official support for UNIP and its policies. When the Rhodesian border was closed in 1973, Congress not only supported the decision, but also called on workers not to be provoked into industrial action.[5] Nevertheless, within this framework of loyalty to Party and Government, the ZCTU sought to influence its policy and practice. The ZCTU treated Humanism as a national ideology open to interpretation, rather than the personal possession of President Kaunda. In 1973, Congress President Newstead Zimba cited Humanism to argue that lucrative travel allowances paid to senior Government officials should be diverted to increase the salaries of low paid workers.[6] Zimba also attacked the slow pace of Zambianisation and 'lazy' political leaders who sought to scapegoat workers for the economic problems for which they were responsible.[7]

The ZCTU leaders, in contrast to their predecessors, provided a consistent public critique of UNIP economic policies. They demanded an incomes and prices policy that would ensure that wages kept pace with the cost of living, and requested the participation of the labour movement in national decision-making.[8] In 1973-74, Zambia's first IMF agreement required the removal of some food subsidies (although this was not publicly acknowledged). There was public uproar at the resultant price increases and Zimba declared that the ZCTU would support demands for compensatory wage hikes.[9] The price rises were revoked by the Government without explanation.[10] It was to be the first of many confrontations between Congress, and the Government and IMF.

In 1974, Zimba was elected ZCTU General Secretary. Frederick Chiluba, President of the National Union of Building, Engineering and General Workers' Union (NUBEGW), was elected President, defeating MUZ's David Mwila, who was characterised by his opponents as a Government stooge.[11] Ongoing tension between Congress and the MUZ reflected discontent amongst MUZ leaders that Zambia's largest and most powerful union was not adequately represented in the labour movement's leadership.[12] As the ZCTU emerged as the leading critic of UNIP policy, it highlighted the comparative moderation of Mwila's leadership. For UNIP, the loyalty of the MUZ leadership was crucial, because of the centrality of the mining industry to the national economy. MUZ Head Office officials had frequent and direct access to President Kaunda, a fact resented by the ZCTU leaders. This access did not however mean that the aspirations of MUZ members were addressed. Disillusionment amongst local MUZ officials with the practical results of this 'preferential' access led them to develop informal ties to the increasingly prominent ZCTU leadership.

These links were facilitated by improved Congress organisation, funded by increased affiliate subscriptions imposed by the IRA. There was a significant expansion in workers' education. A monthly journal, *Workers' Voice*, was periodically published from 1973. New research capacity enabled Congress to respond to UNIP policy announcements. Improved communication and organisation reinforced the ZCTU's claim to represent all union members, whose number increased to 367,000 by 1980, twice UNIP's membership.[13]

At its 1974 Conference, Zimba demanded that the ZCTU should be involved in economic planning and called on Government to ensure the more equitable sharing of national wealth. When prices of essential goods rose in 1975, Zimba called for a national convention of Government, businesses and trade unions to address Zambia's economic crisis.[14] Kaunda claimed that, as members of the National Council, ZCTU leaders were bound by its decision to reduce subsidies.[15] The dilemma of participation was thrown into sharp relief. As Akwetey puts it:

> The National Council was used mainly as the platform for the announcement of ... new policy initiatives by the President...It neither served as a forum for debating public policy nor even putting forward alternatives. Yet membership of this body entailed clear obligations towards UNIP...[16]

The ZCTU leadership's attempt to play a consciously participatory role in a non-pluralist one-party state consistently threw up such contradictions. As Akwetey describes, they resisted direct control under the guise of corporatist participation, but at the same time aimed to shape UNIP's attempts at incorporation to ensure substantive representation in policy-making bodies.[17] Zimba addressed these contradictions at the 1974 MUZ conference. Citing UNIP's policy of 'industrial participatory democracy', Zimba argued that trade unions had a legitimate right

> to have a view about the functions and activities of management be they parastatal or private ... the whole range of industrial and economic issues ... are not decisions that can be left to governments alone ... The greater [Government's] capacity to make massive and major decisions affecting our industrial society, the more crucial it is that we should find ways of ensuring that people's views are properly expressed.[18]

A 1976 ZCTU symposium declared:

> Social progress can only be made if conservative and reactionary elements, whether nationally or internationally based, are countered by strong, independent and representative organisations of workers dedicated to improving their members' living standards and the national wellbeing ... some governments are unfortunately under the illusion that restrictions on the right of working people to organise themselves ... can in some way make development easier to obtain...In fact ... without a free and active trade union

movement, governments will find it harder, not easier, to gain the acceptance of working people of the short-term sacrifices which longer term progress may require.[19]

UNIP did not share this vision. Having enabled stronger central union organisation through the 1971 Industrial Relations Act, it now sought to undermine it at the workplace by the creation of rival bodies. In 1974, UNIP decreed that Party Committees should be established at all places of work, which would:

(a) ensure the strict enforcement of party discipline at work;
(b) ensure that the party programme and policies are being followed and implemented by the enterprise;
(c) see that all people work hard and maintain the highest level of productivity;
(d) instil political and security education in the workers;
(e) counter subversive activities and prevent sabotage in any shape or form.[20]

Party Committees were established, but were not noticeably effective in fulfilling their objectives and were regarded by some mineworkers as workplace spies.[21]

The politics of the ZCTU

The ZCTU's oppositional policies led to it being characterised as right-wing or pro-capitalist, for example in Liatto's depiction of 'Chiluba's right wing economistic trade unionism'.[22] Gertzel argued that, 'They approached economic policy as managers rather than as socialists, preferring the development of an efficient mixed economy rather than an inefficient state capitalism, if that would ensure greater employment opportunities for Zambians.'[23] Such characterisations assume that UNIP's 'state capitalism' represented some form of progressive or 'socialist' system. Setting such assumptions aside, an analysis of the positions adopted by the ZCTU in the 1970s demonstrates that it most commonly adopted a leftist or populist critique of the gulf between UNIP's socialist rhetoric and Zambian reality. Congress criticised the unaccountability of international capitalism and its influence in the Zambian economy and advocated the redistribution of wealth from a bureaucratic elite to the urban *and* rural poor. In 1976, the ZCTU offered this analysis of multinational companies:

> These companies owe no allegiance to any nation or state. They mostly seek to escape any form of democratic control or social responsibility ... Some of their financial and trading practices have undoubtedly had serious repercussions on the implementation of the policies of many governments ... Some of these companies have, moreover, used their enormous resources to interfere in the internal affairs of countries in which they have established subsidiaries...[24]

Congress argued that UNIP's policies must be backed by adequate funding if they were to succeed, calling in 1977 for financial support for those willing to return to rural areas and take up farming. On May Day that year, Chiluba warned that claims that Zambia was a classless society would remain a myth unless a comprehensive social security system was introduced.[25]

The ZCTU was social democratic rather than 'pro-business'; the key criteria for its policy was not an efficient economy *per se* but the welfare of its members and the poor. Thus in 1977, Congress argued that there should be free competition in pricing, to avoid apparently arbitrary price increases that caused suffering amongst the poor.[26] In 1978, the ZCTU complained:

> As the economic crisis deepens the weakness and failures of the Party and its Government's own polices are becoming more and more apparent ... disparity continues to exist between the high earning wage employees and the low wage earning workers. The majority of the working Zambians are still under-paid compared to the earnings of the expatriates ... why has it not been possible for the Party and its Government to ... create uniformity in wages and salaries...?[27]

The breadth of ZCTU criticism widened in response to increasing economic difficulties. Chiluba criticised media 'self-censorship', which he argued hid the extent of Zambia's economic problems.[28] In 1979, he contrasted the neglect of agricultural development to the construction of 'skyscrapers' in Lusaka.[29] When UNIP proposed the forcible deportation of the unemployed to rural areas, Zimba claimed this breached ILO Convention 105 guaranteeing freedom of movement.[30] The ZCTU criticised the employment of unproductive and corrupt bureaucrats in parastatals and Government. Such messages resonated with the majority of Zambians whose living standards were declining. In the rising economic crisis, as (despite its rhetoric) the UNIP Government continued to cooperate with multinational mine capital and as its policies were increasingly defined by the IMF's market conditionalities, the union movement became the focus of resistance to the practical impact of its policies.

Participation in practice - Works Councils

Aspirations for worker participation in decision-making were supposedly to be addressed by the introduction of Works Councils (see Chapter Three). Whilst the IRA was implemented in 1974, Works Councils were not. Chiluba criticised the delay, claiming that 'decision-making remains a paper slogan aimed at defusing unions...'[31]

Business lobbying achieved a further reduction in the powers of Works Councils via a Parliamentary amendment in 1976. [32] Formation of Councils began that year, with the mines amongst the first to introduce them. MUZ National Chairman Mwila argued that worker participation imposed new responsibilities on miners:

> Under capitalism a worker who steals, wastes, or does not promote productivity is doing it with a good conscience because it ... belongs to the enemy who does not belong to his class. But if he does the same in the social order where participation is 100%, then he is doing harm to his fellow workers...[33]

Mwila's rationale depended on the reality of participation. In practice, there was great confusion over the actual powers of Works Councils.[34] In a test case at the Industrial Relations Court in 1978, Luanshya-Roan Mine Works Council was found to have no power to challenge the appointment of an expatriate, because its remit covered only those (Zambian) workers eligible for Council membership.[35] The contrast between such restrictions and the claims initially made for worker participation is striking. Fincham and Zulu's study of Works Councils found that few workers saw them as significant and that most continued to see trade unions as their primary workplace representative.[36] A 1977 Government survey found widespread criticism of 'participation' in the mines amongst Works Councillors and MUZ branch officials. Works Councils did not undermine union organisation in practice, because they did not deliver genuine participation in decision-making.

The ZCTU rejected UNIP proposals for direct worker management and share ownership of companies in 1976-77, warning that this could encourage capitalist inclinations amongst workers.[37] MUZ's David Mwila expressed scepticism, arguing that a worker's 'labour [is] … enough to give him a say in the enterprise.'[38] Business representatives also criticised the proposals.[39] This was one of the first examples of a shared critique of UNIP policy by unions and businesses, which became increasingly common in the 1980s (see Chapter Five).

Unity and dissent in MUZ, 1973 - 1976

Following intra-MUZ conflicts in 1970-71 (see Chapter Three), the union's constitution was amended to increase head office authority. Previously annual conferences became biennial and the office period of national leaders was extended.[40] Departmental officials became members of the Central Executive Committee (CEC), decreasing the proportion of elected members.[41] Criticism of the leadership was tempered by the more repressive political environment, but also MUZ's initial effectiveness in securing pay increases. Interviewees criticise low wages and poor housing, but also recall that wages kept pace with the cost of living and that jobs were secure.[42] MUZ's perceived effectiveness was based on good branch organisation, enabling grievances to be communicated via increasingly well trained shop stewards to the leadership. Leaders briefed public meetings and through communication and relative accountability, members were persuaded to accept unpopular agreements.[43] Consultation offset division and unity was strengthened by miners' lack of alternative vehicles for the expression of grievances under the one-party state.[44] The recent experience of state repression restrained public discontent: one Chingola miner wrote, 'We are afraid to tell you all the mistakes the companies are making in case we find ourselves where Albert Supaile is now.' (see Chapter Three).[45]

Criticisms were however expressed, echoing those recorded by Burawoy.[46] One Luanshya mineworker wrote:

> You Mr Mwila should know that we are already making enough contributions including the one for your luxurious office, the big car you drive in, the mansion that you live in, plus the fat cheque you draw every month at the expense of our little earnings. You enjoy all those for the job undone.[47]

Criticism increased when David Mwila was elected as UNIP MP for Nkana in 1973. Nkana branch officials who called for him to stand down from MUZ were removed from their positions.[48] Konkola miners subsequently threatened to boycott the 1974 branch elections if they were supervised by head office officials who, Branch Chairman Palakasa Chiwaya claimed, had a set of 'blue-eyed boys' they wanted elected.[49] Many officials with a record of criticising the MUZ and UNIP leaderships were nevertheless re-elected, including Chiwaya.[50] Outspoken former detainee Roy Yumbe was controversially re-elected as Nkana Branch Secretary.[51]

Meanwhile, MUZ Deputy National Chairman Alexander Kamalondo alleged that Luanshya branch officials Fidelis Mwamba and Kossam Milambo were plotting to 'overthrow' Mwila.[52] Mufulira branch officials claimed their challengers were reviving the UPP. Days before the Mufulira branch election, eleven candidates were detained by the police.[53] Mwila stated that, 'If some members try to be subversive, there is nothing the union can do.'[54] All were released a few days later, with no explanation; eight of them later successfully sued MUZ for providing false information to the authorities.[55] One of them, Alfred Mukuka, said he had been questioned about the UPP, but said, 'I still do not understand how union elections were combined with anti-Government activities'.[56] If Mukuka was unclear about the link, the Government was not. Copperbelt Member of the UNIP Central Committee (MCC) Alex Shapi secretly wrote to President Kaunda following the detentions, demonstrating the importance of ensuring the political loyalty of the mine townships and the union:

> There is one area where the Party is terribly weak … this is in the Mine Township Compounds. My meetings with the Leaders in the Registered Constituencies, Branches and Sections of the Party were almost a failure … The Leaders seem to be very un concerned [sic] and independent.
>
> Another list enclosed … is the list of names of Ex-UPP previously detained but promised to be Royal [sic – loyal] to Your Excellency, the Party and the Government when they were realised. [sic – released] These are the same people suspected to have resumed their activities against the party and Government.
>
> The situation is being watched by the Party and the Government. Some of the recently Ex-detained Union Candidates in the Mufulira

> M.U.Z. Branch elections are planning to take the Government to court. The other places of concern…is the Top M.U.Z. Conference … The Party and Government should not take the M.U.Z. Elections lightly for the sake of Industrial peace and stability in the country.[57]

Kossam Milambo withdrew a planned challenge to Mwila at this Conference and the leadership was re-elected. At the Conference, Mwila condemned unconstitutional strikes and argued against the 'isolation' of the Union from 'constructive' political activities.[58] Conference resolutions combined loyal declarations with requests for policy changes:

> - We will strive hard to maintain peace and stability in the Mining Industry … We shall do all in our power to keep up the high standard of Industrial Relations.
> - We ask the Government to consider the lifting of the 5% wage freeze imposed in 1969 in view of the prevailing inflationary tendencies in the country.
> - We welcome the introduction of Industrial Relations Act No. 36 of 1971…But wish to make a strong reservation on … the imposition of the removal of strikes action by declaration of the Mining Industry as a whole as essential services. We strongly urge the Ministry of Labour to quickly look into this matter with an eventual repealing of the section.[59]

Mwila criticised the mine companies, but opposed any industrial action that might seek to further these objectives.

MUZ's limited democracy restricted the expression of discontent. The selection criteria for conference delegates and for membership of the Supreme Council were opaque and dependent on Head Office endorsement.[60] Fidelis Mwamba, MUZ Treasurer in the mid 1970s, claims that votes were bought and that dissidents were prevented from attending meetings.[61] Branches frequently called for the direct election of national officials by all miners and other democratic reforms.[62] Despite these restrictions, branches and members exercised significant leverage over the leadership by the use and threat of industrial action.

UNIP's preoccupation with the loyalty of MUZ persisted. After Nchanga Branch Chairman Maxwell Kalesha publicly contrasted his directly elected legitimacy with that of the Presidentially-appointed local District Governor, he was asked to sign a declaration of loyalty to the

Party, but refused to do so.[63] Kalesha claims MUZ opposition to the one-party state was consistently expressed in national and branch meetings.[64] This is not reflected in the minutes of these meetings, and whether such private opposition amounts in practice to public quiescence is open to interpretation. In public, MUZ declared support for the one-party state and called on members to vote for President Kaunda, whilst occasionally criticising specific policies.

Ordinary miners also challenged the extent of party authority. When UNIP's Central Committee declared that it would vet candidates for union office, MUZ leaders accepted the idea, but it was unanimously rejected at an Nkana branch meeting:

> Mr John Chansa said it would be "very unfair" … because miners were "mature" enough to know the right people to represent their interests … Mr Damson Mubanga and Mr Bernard Ngulube … argued that the new system would greatly "fall short" of democracy in the labour movement.[65]

Zambianisation and nationalisation

By 1980, expatriates represented only 4.7% of the workforce, down from 16% in 1964. Complaints nevertheless continued about racist supervisors, for example by stores workers in Luanshya:

> Zambian people got their Independence which was a clear indication to say, they were mature enough to rule and control themselves in every field of man's dwelling and enjoy the fruits of their country. Not long ago, a so called white man came back … to maintain his sort of colonial system of ruling the owner of the country. These things are happening in … the Mining Companies.[66]

Criticism was made of expatriate pay increases in 1975. Minister of Mines Axon Soko justified these as necessary to retain scarce expertise and revealed the contradictions of Zambianisation in arguing that the process 'must depend largely on us retaining expatriates to train and pass on their experience and skills.'[67] As Daniel demonstrates, the real wage differential between Zambians and expatriates was about the same in 1975 as in 1962.[68] MUZ leaders argued that faster Zambianisation would reduce costs by removing expensive expatriates. The pay award led to strike threats, relieved only by the unexpected announcement of a cost of living increase for Zambian miners of 7-20%.[69] Zambians now earned between half and two-thirds of the wages of expatriates doing the same job; the latter also received additional benefits.[70]

In 1976, Rokana laboratory technicians claimed unqualified expatriates were getting K500 per month whilst they received K96. In an anonymous letter signed by 'Mac Jones', they warned:

> It is time when the revolutionary struggle of the working class and working people gained strength (in this capitalistic department) for freedom against oppression and exploitation by capital, because without the black man's labour here the whole plan can collapse within hours of striking.[71]

Few miners believed that mine nationalisation had a positive impact. Mijere's 1980 survey found that 97-98% of miners believed nationalisation was the cause of their hardships.[72] Some respondents reported that nationalisation made it harder to gain adequate pay increases:

> it really became worse. When … the Union went for negotiations. They couldn't get what they wanted … it was so difficult to negotiate with the Government. The Government also defended … the other shareholder who had … 49%.[73]

Nationalisation also led to the appointment of local UNIP officials to Zambianised positions.[74] Yumbe argues that the substantial benefits miners expected from nationalisation went only to 'those people in the top positions.'[75] Others claimed that nationalisation caused a decline in investment, because the Government retained profits for unproductive non-mining expenditure.[76] Evidence suggests, however, that direct Government intervention in mine management was limited in the 1970s.[77] Interviewees who praised nationalisation did so on the basis that it had a limited impact on the industry.[78]

Pay negotiations, 1973 – 1976

Mining industry pay negotiations generally took place without recourse to strikes. The mine companies and the state appear to have sought to protect the MUZ leadership's credibility and reward its political loyalty by the prompt agreement of acceptable pay awards. In the 1973 pay negotiations, MUZ leaders, utilising an uncharacteristically aristocratic approach, argued that the hard work of miners should be rewarded:

> the least [sic] two years had seen a great improvement in industrial relations, and in communications ... there had been large increases in the cost of living ... members were working for an industry which was vital to the economy they should be well remunerated ... It was the labour of the miners which provided Government with the necessary money.[79]

The resulting agreement awarded a cumulative 20% pay increase over three years, staying within the 5% annual pay limit. The unpopular withholding of two days' sick leave was removed and a new social security scheme was introduced. The agreement condemned unauthorised industrial action, and required that

> In the event ... that there is any unauthorised strike or unconstitutional action, the Head Office officials, local officials of the Union and shop stewards shall:
> (1) Promptly order Union members back to work.
> (2) Refuse to aid or assist in any way such unconstitutional action.
> (3) In good faith use every reasonable effort to terminate such unconstitutional action.
> (4) Support the imposition of sanctions on employees who involve themselves in unconstitutional action.[80]

Management was empowered to withhold increments from workers organising industrial action.[81] The agreement was hailed as an example of a new responsible industrial relations climate.

In subsequent negotiations, MUZ challenged the supposed privileged status of mineworkers.[82] The union argued that a contented workforce would be more productive:

> The workers in the industry had done more than their share to contribute to the wellbeing of the rural community, and very large sums of money stemming from the efforts of workers had been invested in the rural areas without apparent beneficial results.[83]

The companies' repeated referral to the national interest to justify pay restraint was now reinforced with reference to state control:

> The Government now owned 51% of the industry, and received taxation on any profits made by the independent shareholders. To

divert any additional money into the hands of employees would be to the detriment of the nation as a whole.[84]

Management cited the declining copper price and resultant economic difficulties to justify pay restraint, and expressed the hope that the union would help explain unpopular decisions to their members.

Despite the low level of strikes, the threat of unofficial action provided an important pressure on the union leadership. In 1976, Mwila urged miners not to strike during forthcoming negotiations.[85] Following the appointment of a conciliator, agreement was reached for a 7% increase for higher grades and 15% for lower categories, over three years.

Strikes under the one-party state

1973 was one of the least strike-prone years for decades, with only 920 hours lost.[86] MUZ was praised for its effective control over wildcat strikes. Small local strikes nevertheless did occur in subsequent years. These were commonly organised by sections of workers at particular mines and, under pressure from company and MUZ officials, lasted only days or hours. Strike leaders were disciplined and frequently dismissed and branch officials and stewards faced the threat of arrest. Local officials often condemned industrial action whilst endorsing the underlying grievance. Reporting the threatened action to management, they would absolve themselves of responsibility for ending it.

The repression of official industrial action led mineworkers to organise and enforce the withdrawal of labour in covert and creative ways. Strikes were organised in secret meetings in the 'bush' and spread through anonymous circulars and posters.[87] Strikers avoided assembling and, fearing victimisation, refused to elect spokespersons. Tactics used to enforce and widen the impact of strikes included the targeting of the Cap Lamp Room. Miners could not go underground without cap lamps, so preventing their issue disrupted production.[88] Strikes were discussed and organised in the lift shafts that took miners to work:

> when you are going underground, you use the cages. Then somebody from the dark would simply start talking. So when you go in sections, that cage which was debating the issue, will scatter, and will discuss with their friends. So the following day, you'd find that people, just refusing to go in to the gate.[89]

In 1975, Wusakile-Chamboli MP Steven Malama claimed that because miners had no legitimate channel for their grievances, they "were slowly reverting to "Katilungu unionism" because they realised that during the Katilungu days they had strike action as a weapon to air their grievances."[90] The expression of grievances was partially displaced from industrial to community settings, aided by the spatial continuity between mines and mine townships. Whilst few strikes achieved their demands, they demonstrated the continued capacity for local organisation that would be demonstrated in the strikes of 1981 (see below).

The threat of strikes influenced the positions adopted by union and companies in negotiations, for example, over the provision of transportation in 1975.[91] In such disputes, tension constantly arose between mineworkers seeking to improve pay and conditions or challenge supervisory authority and their union that, under managerial and political pressure, sought to end such action. Sam Kangwa recalls that union officials

> … were in a very peculiar situation, where they were dealing with their fellow Zambian managers, and at the same time they were trying to please the people who voted them in to power, the workers. So they were torn apart … we used to refuse most of the offers we were being given … the managers would explain to them that production has gone down, and we cannot pay … good salaries. And of course the branch officials, to explain that to us, was very difficult.[92]

During a strike at Broken Hill, local MUZ leaders criticised management for paying low wages and thus provoking the action.[93] MUZ National Chairman Mwila appealed for them to return to work:

> Waving fists and shouting, the workers told Mr Mwila…that they would not resume work until their demands … were met. Mr Mwila … warned that if they failed to report for work today the union would refuse to handle their case. "I urge you to go back to work tomorrow, or even today and give us a chance to sort out things. Leave everything in our hands." But his plea was met with shouts and boos.[94]

As economic conditions deteriorated in the late 1970s, strikes became more frequent, more violent, and increasingly likely to lead to dismissals.

Expressions of membership discontent, for example by Winding Engine Drivers (WEDs), ensured the discussion of their grievances; MUZ negotiators referred to

> … the militant tone of the letter recently received from the Winding Engine Drivers demanding that their rates of pay should be increased to the level of their expatriate predecessors … the Winding Engine Drivers had posed problems to both the Union and the Companies over many years…[95]

This issue led in 1975 to a Copperbelt-wide strike by WEDs which, by preventing miners going underground, had the capacity to halt production. WEDs passed a vote of no confidence in Mwila.[96] The drivers were sent personalised letters threatening them with dismissal, but most could not be delivered because the strikers stayed away from their homes during the strike.[97] No improvements were achieved. In 1977, WEDs expressed a popular alternative definition of Zambianisation, declaring that 'no Zambianisation has taken place in the department because had this been so, the Zambians who took over from expatriates would have taken over their salaries as well.'[98] Drivers were still demanding an upgrade in 1979 when the Minister of Mines admitted that, whilst the principle of equal pay for equal work had not been abandoned, Zambia had to pay internationally competitive wages to attract skilled expatriates.[99]

Public meetings and miners' wives

As noted in Chapter Three, union 'public' meetings provided an important forum for MUZ members to hold leaders to account. Such meetings had to comply with the conditions set out in police permit Form 31. Only those named on the permit could speak from the platform, and only on the subject permitted. Other conditions applied:

> - Every speaker shall … speak into the microphone provided for the recording of speeches and shall cease when requested to do so by the police…
> - No person or persons shall be permitted to lead the assembly … in the organised shouting of slogans or other utterances likely in the circumstances prevailing at the time to disturb the public tranquillity or lead to a breach of the peace.
> - The applicant shall immediately cancel the assembly … if, in the opinion of the said police officer, the remarks, speeches, songs or

> other matters are likely to lead to or encourage a breach of the peace or any other unlawful act.[100]

Miners and their families attended public meetings in their thousands; they provided an opportunity to hear directly about negotiations, and to put questions and criticisms to head office officials.[101] Wives attended the meetings for a number of reasons. Enock Chansa suggested that they needed to find out their husband's salary from the union, rather than relying on him to report it honestly.[102] George Mukoma:

> … when the wives start, they are almost the first complainants. They are the people who realise you don't have enough money in the home … that's why they … become … interested in attending these meetings to go and hear what the husbands will have.[103]

Wives were also concerned with the provision of township services, but lacked formal representation with the companies that provided them.[104] On occasion, they took overt action to represent their discontents to the mine companies, and to MUZ, which negotiated over their concerns, but in which they were not represented. In 1978, women in Luanshya demonstrated over the closure of a township clinic. The Assistant Personnel Superintendent expressed his discomfort at finding a demonstration of 200 women outside the General Office:

> He told MUZ Branch Officials that such demonstrations should not be encouraged when there was a proper laid down procedure … such demonstrations were indications of poor leadership in the Union … MUZ replied by endorsing [his] remarks, but explained that they did not incite the women to come and demonstrate … they have urged the women to be patient … whilst MUZ was still discussing their problem…[105]

Six women representatives were invited into the meeting, where the Superintendent

> … warned that demonstrations were illegal … He advised the women to air their problems through MUZ, the body recognised for the representation of employees' interests … The women representatives complained that they brought their problems to the attention of MUZ last year and have had nothing worth while [sic]

> forthcoming their way … they had felt ignored over this problem, whilst their suffering and hardships mounted … The women requested the [Superintendent] to explain to them the circumstances that necessitated the closure of the Clinic.
>
> The [Superintendent] told them that he could not discuss Management's reasons for closing the Clinic because the company does not recognise dependants as representing employees on any matters, however, he informed them that MUZ was fully aware of the reasons that necessitated the closure of the Clinic and it was up to MUZ to brief them.[106]

The MUZ representatives then persuaded the women to disperse. This episode suggests the difficulties faced by both companies and union in their attempts to confine themselves to 'normal' industrial relations, in the face of women's expression of their grievances. However, as Rodah Matoba, one of the first female MUZ office holders describes, the official invisibility of women also had its advantages when it came to such demonstrations:

> … that woman is not recognised by the Company. And even children … All these are dependents. Even if they're to cry out, Management would not look at them, because they don't know them. When you fight, nobody will apprehend you. Nobody will charge you for anything.[107]

Economic decline and rising conflict, 1978-81

1979 saw a substantial increase in industrial action in Zambia. Major strikes took place on the TAZARA railway; on the Nakambala sugar estates, during which cane fields were burnt and the police fired on strikers; and amongst teachers and junior doctors.[108] The *Times of Zambia* noted the evident disunity between unions and UNIP and warned that (at a time of Rhodesian military attacks) external enemies could take advantage.[109]

The 1978 ZCTU conference called for an end to the 5% wage limit required by the IMF.[110] Conflict over delays to the proposed wages and incomes policy led Congress to demand an end to Government's role in wage setting. The ZCTU General Council resolved that workers should no longer be treated as 'mere factors of production but equal participants to the formulation of national policies'.[111] The 5% wage limit was lifted and collective bargaining introduced in June 1980. This was a significant victory for Congress lobbying, enabling the union movement to potentially gain increases for members through their own initiative.

However, this success was undermined by the removal, under the IMF programme, of subsidies to all foodstuffs except mealie meal. Threats of a nationwide strike against the price increases were made and Congress demanded a general 30% pay rise.[112] When Minister of Labour Josiah Lumina claimed that the strike threat was politically motivated, Chiluba, who consistently defended the labour movement against accusations of politicisation, asked rhetorically, 'Who told the minister that a strike action is political and not industrial? Is copper political?'[113] The *Times of Zambia* argued that, 'Part of the function of the ZCTU leaders is to ... stop strikes, not to inspire them ... It is quite clear that the ZCTU leadership is using industrial disputes for purely political purposes to try to deliberately embarrass the Party and its Government. Mr Chiluba and his ZCTU colleagues must know that this is not Poland.'[114]

The Copperbelt witnessed a violent police crackdown in which thousands of workers breaking an unofficial curfew were arrested. The UNIP Youth League was directed to look for those trying to discredit UNIP and its Government, specifically including unionists.[115] Kaunda publicly linked the coup plot revealed in October 1980 (see Chapter Two) to the ZCTU's call for strike action. [116] Despite ZCTU denials of involvement, UNIP investigated 'foreign infiltration' into the union movement, and Congress leaders had their passports withdrawn.[117]

As UNIP advocated scientific socialism whilst implementing aspects of economic liberalisation, Chiluba told the ZCTU's General Council in 1981 that scientific socialism could not work in Zambia because a lot of people had amassed wealth in the form of property and businesses. To loud applause he declared, 'When our leaders show us honestly that we are going socialist and all our families will not be allowed to hold property on our behalf I promise this nation that this congress will talk, eat and walk socialist.' In the meantime the ZCTU, he promised, would not be silenced.[118]

By the late 1970s, economic decline forced the MUZ leadership to be increasingly critical of the failure of both Government and the mining companies to meet their members' demands. Whilst Mwila maintained his public loyalty to the Party and its Government, he argued that mineworkers were making a disproportionate sacrifice in efforts to rehabilitate the industry.[119] In 1978, Mwila told the Supreme Council that Zambia's

> … was a sick economy. He said that the workers were not to blame for this state [of] affairs … Mineworkers have responded favourably well to the cost saving exercise … workers' patience was running out…the present cost saving measures … seemed to be aimed at local employees and not expatriates … if this situation did not change MUZ would not accept any disciplinary measures taken against its members for any unconstitutional work stoppages…[120]

At the 1978 MUZ Conference, Mwila again defeated a challenge from branches calling for 'new blood' to replace his 'weak leadership'.[121] A Conference report calculated that real wages had fallen by 24% from 1970 to 1976.[122] Whilst Mwila assured Kaunda of the union's support, he also complained that

> … criticism of the leadership is regarded as counter revolutionary … Democracy denotes a Government of the people by the people. The Party and its Government is for the people and if that is true … then constructive criticism should be allowed…[123]

Mwila also expressed frustration at the continued closure of the Rhodesia trade route:

> Workers and the Mining Industry will for many more years to come continue to be the nucleus of our economic well being. As such the Industry must import machinery and spare parts cheaply using the cheapest and shortest routes … I know I am going to be called names by those who pretend to be patriots as a counter revolutionary, but I know I am expressing the wishes of those I lead … Ideologies are for politicians but can never feed the Nation … We must bring pressure on our leaders to ease the sufferings.[124]

Simon Kapwepwe had expressed similar views that year. In this tense period, Mwila was briefly detained over his unauthorised possession of UNIP Central Committee minutes. At a secret State House meeting with UNIP leaders, Mwila criticised

> … the way he was treated by the Police … the MUZ Executive had met on 16/9/78 to discuss the allegation that miners were supporting dissidents. The allegation was contained in a copy of the

> Central Committee minutes which had come into his possession to the effect that Mr. Kapwepwe had chosen the Copperbelt because of the support he had from the labour movement particularly from MUZ.[125]

Mwila expressed the union leadership's anxieties regarding its deteriorating relationship with its members:

> ... miners had played a significant part in the freedom struggle and were a very sensitive people. As they worked, lived and patronised the same places together, they had ample time to discuss issues and were bound to react unfavourably to any situation which displeased them ... MUZ leaders and those in the labour movement as a whole had lost some popularity among their members for expounding Party and Government policies. Some members were advocating for a militant leadership ... it was embarrassing for his organisation to be suspect at a time when union leaders faced tremendous difficulties in explaining to their members about shortages of essential commodities and the economic problems the country was experiencing.[126]

Mwila stood down from Parliament at the 1978 elections, but it is unclear if this resulted from his perceived disloyalty to UNIP, pressure from within MUZ, or disillusionment with the political process. Mwila was now frequently absent at international meetings. His Deputy, Timothy Walamba, increasingly became the public face of a more critical MUZ. It was in this context that the 1979 pay negotiations took place. MUZ again criticised the payment of merit-based increases to expatriates, complaining that "they were faced with the problem of convincing their members of the justification for a far greater increase ... which had been given to expatriates alone."[127]

It is difficult to definitively assess wage value against real inflation. The Government pay limit was partly compensated for by the concession of a range of allowances. Official inflation figures do not reflect the higher prices paid on the black market that was now a major source of essential commodities. However, an attempt in 1979 by the companies to explain low output provides some insight:

> This is attributed principally to the low morale of the workforce and resulting low efforts, the loss of skilled expatriate labour, shortages

> and supply problems ... Basic salaries have "slipped" since 1976 so that the average employee has experienced a 30 to 35% reduction in the purchasing power of his income, resulting in reduction of standard of living.[128]

The companies were in favour of a real-terms pay increase to boost production and in 1979 were temporarily exempted from national pay limits. Inflation was now rising so rapidly that the triennial pay round was abandoned. In pay negotiations in 1980, MUZ opposed Government pay controls and for the first time suggested that unless conditions were improved, an official strike might be called.[129] Walamba was in the forefront of this more militant approach, telling miners in Chambishi, 'We leaders in MUZ are prepared to go to jail.'[130] Company officials took into account the danger of industrial action, privately advising that ' "ability to pay" ... must be balanced against need to pay: a long and/or total strike can be most damaging.'[131]

The 1981 settlement included a review of the unpopular job evaluation agreement and a new contributory pension scheme. The new wage levels were not initially disclosed and when they were briefed about the agreement, miners reacted angrily. In Nchanga, Branch Chairman Kalesha was physically attacked. Charity M'tonga recalls:

> What miners wanted from Mr Kalesha [was] to tell them to go on strike ... They were forcing him to say 'Tell us! Say that you must go on strike tomorrow!' So Mr Kalesha couldn't say that, you see? ... they lifted him ... and said 'Can you say that ... we are going on strike tomorrow? If you won't say we'll drop you down!' It was terrible until the policemen came to his rescue.[132]

The local MUZ office was burnt down following the abandonment of the meeting.[133] Kalesha, however, prevented the police arresting any of his members, insisting that branch leaders had to be prepared to face the physical challenge of their members.[134]

Company statistics revealed that mineworkers' earnings were now comparable to other workers: a G8 (general workman) earned K90-K170 per month, whilst similar workers in other industries earned K67-K167.[135] These figures did not include non-wage benefits, particularly the provision of township services. It was a threat to these that prompted the most significant confrontation between the Government and the labour movement since Independence.

Decentralisation of local government

The 1980-81 conflict over the decentralisation of local government and the incorporation of mine townships has been analysed in most studies of post-colonial labour relations. Mijere claims that the conflict demonstrated the mineworkers' apolitical and aristocratic nature, selfishly defending the preferential services provided by the mine companies.[136] Burawoy argues, in striking contrast, that the conflict was fundamentally political:

> The most significant feature of these strikes was their explicitly political character, at least in their immediate goal. They were directed at the state in defense of the independence of trade unions and not motivated by short-term economic demands.[137]

The dispute began as a defence of the services provided by the mine companies and the relative political autonomy from the one-party state that this enabled. This brought mineworkers into conflict with UNIP's attempts to subsidise local government with mine company resources and to extend Party authority to the mine townships (an enduring UNIP concern demonstrated in Shapi's letter to Kaunda, above). This led to strike action in protest at attempts by UNIP to punish union leaders for their failure to control their members. This mineworkers' dispute, as was so often the case in post-colonial Zambia, had both economic and political causes and consequences.

The integration of mine townships into Zambian local government had been discussed since Independence. A report by UN advisors to Government in 1968 found that the mine companies welcomed the opportunity to divest themselves of the responsibility for housing and other services.[138] However:

> They were unanimous … that their workers … would resent coming under the jurisdiction of a Local Government Authority rather than the Mining Company … [which was] prepared to attend to grievances promptly because they want to keep the workers happy – otherwise production might suffer.[139]

The report also found that, in most Copperbelt towns, the mine townships represented the majority (58-74%) of the population. Whilst this was reduced by the post-Independence urban influx, the spatial concentration of the workforce was maintained. For example, in 1974 virtually all the Rokana mine workforce of 11,634 lived in just three areas, more than two-thirds in the twin townships of Wusakile and Chamboli, adjacent to the main mine plant.[140] As in the colonial era, there was high continuity between mine workplace and community experiences and between union and township organisation.[141] This did not lead, as Mijere assumes, to a paternalistic attitude amongst mineworkers towards the mine companies. As the 1968 report suggests, mineworkers generally believed that companies had only conceded services under the threat of riots, strikes, and pressure from the mineworkers' union.

Mineworkers (though not their family members) elected township councils, enabling significant local control over services. Charity M'tonga describes her role as a mine township councillor in Nchanga:

> … whenever we had a meeting, we got the computer run off … we would look on them, and see whether the money has been used to intended cost … if the money is not in, we would say [to company managers] why are you taking this money? Why are you not doing this? … We made sure that the town was clean … That was our duty. People would stand and be elected by the whole community.[142]

Unaccountably, Mijere fails to mention the township councils, claiming that 'The inhabitants of the mine townships exercised no political power over the appointments of the members of the Board.'[143] From this false basis, he argues that, 'For the miners the issues of personal responsibility, political self-realization and decentralization were not as important for them as the material benefits accrued from the old company system of administration.'[144]

In 1975, an official working party was established to investigate decentralisation. MUZ initially discussed boycotting the exercise, but ultimately agreed that David Mwila would be a member of the working party.[145] In its submission, MUZ explained that the services mineworkers received were made possible by profits generated by their hazardous labour, and that in the past

> … the Mineworkers had to use the strike weapon in order to achieve their objectives … [some] strikes have been provoked by the Mining Companies' reluctance to offer better social services to their employees. It is therefore a matter of conjecture whether left to themselves the companies would have improved conditions in the townships.[146]

Mufulira Branch Chairman John Manda similarly argued:

> Whatever development you see around the mine townships is blood soaked, by those who lived and emerged from primitive buildings some 40 years ago to modern buildings, as cultural heritage of the present miners, through hard work and devotion to duty.[147]

The MUZ Supreme Council resolved to reject the integration proposals, and warned that miners could

> … resort to strike action, go slows and other actions registering their disapproval … the Industrial unrest we are predicting may be at a scale hitherto unwitnessed in the history of the Mining Industry. As a very responsible organisation with the interests of the nation at heart, we would hate to see such a thing happen.[148]

The working party found mineworkers to be overwhelmingly opposed to integration. At a public meeting in Nchanga, Branch Chairman Kalesha argued that as municipal councils were unable to provide adequate services to non-mine areas, they would not be able to do so for the mine townships.[149] When the working party arrived, more than 10,000 miners assembled and threatened to surrender their UNIP cards in protest at the proposed integration.[150] In Nkana, the working party was locked up by protesting miners, and Mwila was nearly manhandled.[151] A poster displayed in Luanshya declared: 'Go back to Lusaka and tell them to resign and be replaced before things go out from hand … It will be impossible for the local Party officials to control the miners. Please think it over. Away with integration.'[152] Alexander Kamalondo, now the Luanshya District Governor, claimed external enemies were using the mineworkers to disturb the peace.[153] The working party report was not acted upon.

By 1980 the companies, struggling to maintain services to the growing mine township population, sought to persuade miners to send members of their extended families back to rural areas. The Manager of Administration in Luanshya explained:

> The overcrowding in many of the houses was in certain cases due to extended family ties and in other cases due [to] polygamous marriages. The sociological problems brought by such overcrowding

> cannot be overlooked since the facilities provided by the company covered an average [nuclear] family.[154]

The companies' need to reduce costs coincided with UNIP's desire to achieve political control over the mine townships and address financial weaknesses in municipal councils. This resulted in the presentation of the Local Administration Bill to Parliament in August 1980. The Bill proposed the replacement of all existing councils with integrated District Councils. Councillors would be chosen by approximately 60 local UNIP officials (municipal councillors were previously elected by c.2,500 constituents).[155] A District Governor, appointed from outside by President Kaunda, would head each council.

At this time, Mijere surveyed opposition amongst Nkana and Luanshya miners to this proposed 'decentralisation'. As advantages of life in the mine townships, miners cited peace and security, cleanliness, medical services, low rents and the provision of housing near to the workplace.[156] Unlike in municipal areas, houses were repaired, and water and electricity were free. Mijere's questioning revealed a distinctly non-aristocratic response:

> When I asked the respondents why they would fail to pay the [higher] rent [in council areas], they replied that the money they earned was not enough to meet their and their families' needs. One miner said clearly, 'I work only to buy food with the money. I have now stopped buying clothes for myself; I walk in these rags.'[157]

My interviews revealed similar concerns: mineworkers feared that the conditions in mine townships would become as poor as those in municipal areas.[158] The use of mine companies' profits to underwrite local government would also threaten their economic position.[159] Some opposed the policy on the basis that it would extend UNIP's authority over the mine townships, and the miners themselves.[160] Timothy Walamba argues that, 'Decentralisation of local government was political … UNIP wanted to control local governments … a Governor was a political leader … He would want only his colleagues, his Party cadres, to be accommodated.'[161] MPs criticised the proposed legislation for reducing democratic accountability, and some called for the popular election of Governors. One told Parliament, 'In this country we are running away from true democracy … we will go down in history as a group of people who legalised the disenfranchisement of the Zambian people.'[162]

The Bill was deferred by Parliamentary opposition.[163] However, after Kaunda told UNIP's National Council that voting for local government was not a constitutional right, the Party proceeded with preliminary elections, indicating its intention to introduce the new system. In this context and coinciding with tension over pay negotiations, the MUZ Conference in 1980 was markedly militant.[164] Mwila declared that the proposed system would bring chaos:

> ... the Republic Constitution confers upon you and me the democratic rights to vote for our representative...in local government ... Peasants and working people are the vanguard of the Zambian revolution. Working people are organised through the Trade Union movement. These workers movement cannot allow our leaders to pass laws as they like.[165]

The Conference went well beyond mere criticism, in resolving that 'the imposition of this Act will result in Industrial action which will be too ghastly to contemplate ... Those leaders who have participated in these local Government elections should either withdraw their candidature or resign from MUZ leadership.'[166]

Days later, Kaunda suggested that labour leaders were implicated in an alleged coup attempt (see above). Under severe political pressure, MUZ nevertheless implemented the resolution, suspending sixteen shop stewards who had participated in the initial elections. John Kapansa, a UNIP loyalist who nevertheless opposed the new system, was one of them:

> The resolution was ... every miner should not participate in local government election. So when we came, we were not allowed to contest the election. But, while I was there [at the conference], I was elected in absentia ... I came to ask my [MUZ] comrades here, I told them now look, it is important that I can be here as a union official. Because we [are] going to tell the Party and its Government the importance of a worker. [But] on the Union side, they didn't ... see it like that.[167]

Kaunda ordered MUZ to explain the suspensions.[168] Mwila wrote to him privately, denying any link to the coup attempt. He also responded positively to Kaunda's personal assurance that the services mineworkers received would be maintained, and asked for this to be incorporated into the revised Local Administration Bill to be re-presented to Parliament.[169] At a subsequent Supreme Council, Mwila declared that

> … he was placed in a very awkward and difficult position…the Party was very angry with MUZ because of its constant refusal to … have the sacked [sic] Shopstewards reinstated. Comrade Mbao told the National Chairman that as it was not the intention of the members to implicate him (the National Chairman) he should therefore be rest assured of the members' total support for him and that the question of reinstating the sacked Shopstewards as far as the Union was concerned was out he concluded. All the members unanimously agreed with him.[170]

Mwila was thus given the Council's full support, but no room for manoeuvre. When the amended Bill was reintroduced, it specifically declared that all municipal services provided to mineworkers by the mine companies would continue under the new local government system.[171] If the protests against the legislation were a selfish defence of mineworkers' interests, opposition should now have fallen away. However, Mwila and Walamba failed to persuade the Supreme Council to reverse the Conference decision:

> … the National Chairman said that it was not the wish of his team to be in confrontation with the Party … He appealed for patience and earnestly requested the members to adopt a spirit of reconciliation by avoiding direct confrontation with the Party. The Deputy National Chairman warned that the Party and its Government would not sit idle, but would hit hard … to maintain the image and existence of MUZ leadership, it would do a lot of good if the Supreme Council rescinded its decision … Comrade Peter Chewe … strongly appealed to the members to be vigilant; by sticking by their guns. Then the Councillors understood to bear full responsibility of whatever action the Party and its government might choose to take…[172]

In January, after the Local Administration Act became law, seventeen ZCTU (including Chiluba and Zimba) and MUZ leaders were suspended and then expelled from UNIP. The Party's Secretary General explained, 'The ZCTU has virtually ceased to be concerned with labour matters but has become a political organisation.'[173] Kaunda claimed, 'the present MUZ is operating as cover for the UPP.'[174] The UNIP Central Committee declared, 'the expelled leaders will cease to be members of the Party, Government and public institutions to which they have been appointed or elected.'[175] This was interpreted to mean that they would also be removed from their positions as union leaders.

Four days later, all Chililabombwe mineworkers struck to protest against the expulsions.[176] The strike spread to become the first industry-wide action since 1966. Numerous confrontations took place between miners and police, who dispersed meetings with tear gas. Kitwe's Political Secretary was beaten when he sought to persuade miners to return to work and the car of Copperbelt MCC Alex Shapi was stoned. In Kitwe, as an indoor meeting was surrounded by riot police, a large crowd walked out when the national anthem was played. One anonymous circular read:

> All workers out: in defence of trade union rights; against Party dictatorship; human rights violated; leave trade union leaders alone; ... expulsion of union leaders from the Party means expulsion of all workers belong[ing] to trade unions. Workers are one.[177]

Branch officials absolved themselves of responsibility for members' actions, blaming the Party for provoking the issue. Although UNIP accused ZCTU leaders of organising the strike, contemporaneous reports suggest it was organised by rank-and-file MUZ activists.[178] On 23 January, the Ministry of Labour declared that the expelled leaders would retain their union positions and Mwila now called on miners to return to work.[179] On the same day, a full-scale riot took place in Mufulira, with hundreds of miners threatening to beat anyone who returned to work. Walamba's attempt to persuade these workers to return backfired, in a way that revealed mineworkers' enduring suspicion of their leaders. Mineworkers demanded to see union leaders' returned UNIP membership cards and the passports confiscated from them.[180] George Mulenga was present:

> I was a shop steward, and we had a meeting ... where Walamba, he and others came to address workers to go back to work. But when they came they found that people had already gathered stones. They were saying if they don't produce the passports we will stone them. It means they're also sell outs ... But when they heard of that ... they just said oh! Excuse us just for ten minutes we are going back to pick them.[181]

Walamba and other leaders did not return and armed police dispersed the angry crowd.[182] Nordlund rightly concludes that 'these strikes on the Copperbelt appear not to have been organised centrally by the ZaCTU [ZCTU], but were rather proof of the legitimacy that these leaders enjoyed among the membership.'[183] However, this event shows that such legitimacy was contingent on those leaders' actions.

The next day, during disturbances in Kitwe, 14-year old Samson Nkani was shot dead, whilst a baby was burned when a tear gas canister was thrown into a house. Armed police shot and critically injured three Mufulira miners stopping colleagues returning to work.[184] It was later reported that 58 people were wounded during the strike and K20m mine production was lost.[185] Chiluba and Mwila attended Nkani's funeral, calling on MUZ branches to financially support his family's planned legal action against the Government.[186]

Former trade unionist Basil Kabwe was now suddenly appointed Minister of Labour. Kabwe immediately condemned the police handling of the strike and addressed the ZCTU General Council. Chiluba and Walamba, apparently unnerved by the unrest unleashed in their defence, also adopted conciliatory positions; MUZ reinstated the expelled stewards. In February, a government reshuffle saw the Prime Minister and UNIP Secretary General replaced.[187] The new Secretary General, Humphrey Mulemba, pledged that the Party would revamp its tarnished image and toured the Copperbelt to promote reconciliation.[188] Kaunda, however, continued to suggest that the union movement was being used by outside forces.[189] Chiluba insisted that the labour movement would not provide a rubber stamp for UNIP policies.[190]

The mealie meal conflict

The January 1981 strike demonstrated the power of overt industrial action, and two longstanding grievances sparked major protests in the following months. Shortages of mealie meal had been steadily worsening because of controlled prices and the resultant smuggling to Zaire. Mineworkers' long hours made it difficult to locate scarce supplies. In 1979, Luanshya MUZ blamed this for rising absenteeism, and requested that RCM sell mealie meal in the townships.[191] Food shortages were having severe effects on miners at work, as Japhet Sekwila remembers:

> … you would go down underground. Then, just in the cage you'll find somebody fainting. You'll take him to the central place and what is happening is, for the past two days I haven't eaten. Why? …

> My wife has been queuing up for mealie meal. There's no mealie meal.[192]

Miners' wives claimed they were forced to have sex with traders to secure food supplies.[193] Such reports were common in May 1981, when a MUZ Executive meeting heard that most areas were suffering shortages. Chililabombwe, close to Zaire, was most severely affected. Miners' wives there demonstrated at the offices of MUZ, the District Governor and the mine General Manager, threatening to stop their husbands going to work on empty stomachs.[194] Prime Minister Nalumino Mundia promised Walamba, the acting National Chairman, that the situation would soon return to normal:

> … because of these demonstrations, because of these petitions from the Branches, I made an appointment of meeting the Prime Minister … to tell him of exactly what was happening on the Copperbelt ... I said now look, mining is not like clerical work … You can't go underground with an empty tummy … And miners at that time had started collapsing in the cage as they are taken underground … I said 'how long?', the Prime Minister said 'Just give us seven days'. Maize will be provided. I said no problem … I assured my colleagues from all the Branches … maize is going to be provided as soon as possible … That week elapsed, there was no maize forthcoming. Second week, gone, no maize forthcoming. What miners started doing now was to absent themselves from work, to go into rural areas to start looking for maize … As a result of that, some of them were fired as deserters after ten days. I said over my dead body … I addressed a rally in Kalulushi … I told them I had met the Prime Minister…but…maize is…nowhere to be seen. What did you expect of me? So, they just decided there and then to say no, Mr Walamba … the only next step that we can take is to not report for work … this is where now UNIP started becoming unpopular … They no longer cared for a common man.[195]

Walamba felt betrayed by the failure of the Government to keep its promises and effectively abdicated responsibility for what happened next.[196]

In June, the mine companies secured temporary supplies. Meal was provided in Chililabombwe, but then discontinued. MUZ warned management that it would not be held responsible for the consequences.[197] The role of miners' wives was highlighted in a *Times of Zambia* editorial appealing that, 'Our comrades in the mines should therefore put the nation first and not go along with the "petticoat incitement" that they stop work over mealie meal.'[198] On 6 July, it was miners' wives, many wearing their husbands' overalls and hard hats, who mounted pickets to prevent their husbands going to work, physically attacking those who did so.[199] The shops of alleged smugglers were attacked and youths were injured when the police fired tear gas.[200] The action was organised by miners and their wives:

> … we decided to force the Government so that they can give us mealie meal. That's why we decided to bring strikes and civil disobedience … the women take part in that demonstration. They even wear sometimes overalls…and they used to block … roads … No-one was going … to work … When they hold you, they'll beat you up there. They play a very big important [part] in that demonstration because they were very organised.[201]

Miners' fear of dismissal was one factor that led to wives, unrecognised by the mine companies, to play such a prominent role. George Mukoma's reports a decidedly unofficial form of industrial action:

> … trying to report for work, I saw a group of women by the gate … I started going until I was missed by a stone ... I looked at a distance where that woman threw that stone. Then I realised it was a man … not really a woman as such … there were men also in the same gear. But pretending they were women, they had some headdresses and … fabricated breasts and all but they were men … they didn't want to appear like miners themselves … for … fear that they would lose employment … So women did the chanting, but men did the beating … and they closed the whole plant for some days.[202]

Evidence is limited, but it appears that the restrictions imposed on formal industrial activity and its consequent displacement to community settings, enabled wives not only to play a more prominent role than they might otherwise have done, but also to ensure the prioritisation of grievances immediately relevant to them.

The strike spread to Nkana the following day. Whilst national MUZ leaders agreed to urge their members to return to work, Nkana Branch Chairman Moses Bwalya refused to do so and his Chililabombwe counterpart argued that miners had a genuine reason to strike.[203] There, MUZ leaders pleading for a return to work were stoned and chased away by miners' wives. Walamba claimed that smuggling in Chililabombwe took place with full knowledge of UNIP, but denied that he instigated 'the strike as a means of gaining popularity and campaign against the National Chairman in order to oust him.'[204] Minister of Labour Kabwe appealed for a return to work to prevent the collapse of the national economy. The strike ended on 14 July. Subsequently, both mine companies agreed to provide mealie meal to miners on credit. A further strike occurred almost immediately, on another longstanding issue (see below).

The graduates and artisans strike

MUZ was consistently critical of the slow pace of Zambianisation.[205] By 1981, Luanshya's workforce of 9,118 included 161 expatriates and 404 African 'aliens'. However, 100 of 185 senior and management staff were expatriates.[206] Whilst whites had largely been removed from workface supervision, they retained significant control over the running of the nationalised mines. For the mine companies, the replacement of expatriate supervisors with Zambians had, as noted in Chapter Three, created major problems of effective supervision. The mine companies sought to replace the old racial division with a new form of separation, by removing Shift Bosses from MUZ:

> It was considered that front line supervisors (Shift Bosses) should be part of Management. Being represented by the Union subjected them to enormous pressures, both on the job and socially ... Protection of Zambian supervisors was of paramount importance if discipline were to be improved and productivity raised.[207]

Senior and skilled mineworkers aspired to the middle-class lifestyle they associated with their skills and training, but were frustrated by unequal wages and distrusted by those they supervised. Interviewees such as Manjola Banda reported that 'those Zambians who ... were supervisors at that time they'd worked with these people [expatriates], again they continued to use the same language ... So, Zambian supervisors they were abusive.'[208]

Daniel demonstrated that pay agreements in the 1970s had widened the pay gap between skilled and unskilled mineworkers. From this, he speculated that industrial action was most likely to occur amongst unskilled workers. However, skilled mineworkers were frustrated at what they saw as their marginalisation in MUZ.[209] The mines increasingly struggled to retain skilled Zambian workers, as higher union density and effectiveness pushed up wages in other industries. The perennial demand for 'equal pay for equal work' was, for these workers, still a burning issue. In July 1981, in the midst of the mealie meal protests, 90 skilled miners in Luanshya presented the following petition:

> We, the workers, condemn the practice of paying workers (of whatever qualification of skill) on the basis of their race, colour or country of origin, as this is tantamount to practicing an apartheid policy of discrimination and exploitation. We are supposed to be a humanist society.
>
> We demand that all workers doing the same type of work be paid the same amount of money…
>
> We demand that our salaries/wages be brought up to the levels of our expatriate counterparts in all the job categories with immediate effect…
>
> We hereby give the parties concerned (M.U.Z., C.I.S.B., The Ministry of Labour and Social Services) up to 15th July, 1981 to come up with a concrete, positive and satisfactory solution … we, the Workers, have (actually) waited for more than 16 years for a solution to the exploitation of our labour power and the ever rising cost of living that continually keeps us in poverty and misery.
>
> If, by 16th July, 1981, no concrete, positive and satisfactory solution has been arrived at by the parties concerned, then the axe of the working class shall strike at the forces of reaction that are bent on retarding the progress and development of the working class.[210]

The reaction amongst unskilled workers to the strike call was mixed. Interviews suggest that many supported these demands as a matter of justice and solidarity. Some believed that a pay rise for higher grades would further demands for a more general wage increase.[211] Others however opposed the strike, arguing that most ordinary miners did not support the 'selfish' demands of skilled workers.[212] The MUZ leadership had an equally mixed reaction. National Chairman Mwila initially dismissed their complaints. Walamba subsequently made public statements in support of their demands.[213] As the deadline approached, Mwila announced that artisan and graduate miners would get new salaries and appealed for the strike to be called off.[214] However, Rokana mine was paralysed on 17 July, as riotous artisans, graduates and technicians forced and/or persuaded the 13,000-strong workforce to strike.[215] Strikers mobbed Minister Kabwe as he attempted to address them, demanding to know the pay award that had been agreed. The *Times of Zambia* claimed: 'It would seem however that the MUZ leadership has lost control of its members to some unknown forces using rumour and mob psychology to incite the miners.'[216]

The next day, the strike spread to most other mines. Rokana mine was closed, but in Chingola, only 620 artisans, shift and section bosses struck.[217] As ZCTU and MUZ leaders were summoned for talks with the Government, some miners claimed that their leadership had become stooges of the mine companies and that Mwila had lied about the pay rise (he subsequently admitted that no agreement had been reached). By 21 July, half the mines were on strike and 15 arrests had been made. There was a general return to work by 24 July. 590 workers were dismissed, although many were later reinstated. Some unskilled mineworkers now feel that they were misled into participating in a strike that was not in their interests.[218] By the end of 1981 however, the mine companies had agreed to pay Zambian artisan foremen and shift bosses the same basic wage as their expatriate counterparts (expatriates still received additional benefits).[219] The strike demonstrated that, as Nordlund argues, 'it is obvious that the Mineworkers Union did not enjoy the support of its rank-and-file.'[220]

Detentions

Three days after the strike ended, four union leaders were detained: Chiluba, Zimba, ZCTU Assistant General Secretary Chitalu Sampa and MUZ Deputy National Chairman Timothy Walamba (but not Mwila). They were held under the Preservation of Public Security Regulations, under which President Kaunda had the authority to detain anyone he saw as a threat to Zambia's national security. Walamba recalls: 'I was called names … as one of those who were planning to overthrow the Government. Me, when I said these things I was merely talking about the shortages of maize against my people who elected me to the position.'[221] Kaunda claimed that the union leaders had 'sought to use workers to achieve their political aims' and had organised unemployed people to picket, to prevent miners from going to work. He also argued that the 205,681 days lost in wildcat strikes since 1979 represented a threat to law and order.[222] ZCTU Deputy Chairman General Herbert Bweupe advised union members to be calm, but did not rule out a protest strike.[223] In contrast to other political detainees, the union leaders' appeal came to court within two months; there seems little doubt that the threat of renewed industrial action encouraged the prompt examination of the case.

The state claimed that the detainees' objective was to take over national leadership and cited their remarks that, it argued, had caused the strikes.[224] In contrast, the defence case rested on the argument that miners had acted on their own initiative, not on instructions from union leaders. This was accepted by judges who ruled the detentions invalid. On his release, Sampa insisted that the labour movement would continue to criticise Government policies that were against the aspirations of the people.[225]

Conclusion

This chapter has shown that, notwithstanding the decline in industrial action and the lack of open political challenge to UNIP following the declaration of the one-party state, the Copperbelt-based labour movement and the mine townships in particular provided a central locus of resistance to UNIP's attempts to achieve effective control over Zambia's economy and society. UNIP initially sought to address industrial unrest through the incorporation of union leaderships, the banning of strikes, the introduction of Works Councils and the prompt award of annual pay rises. This briefly achieved a reduction in strikes, and it appeared to have delivered, for good or ill, the acquiescence of the labour movement.

It is impossible to know whether, if the copper price had not collapsed in the mid 1970s, this approach to labour relations would have achieved the long-term hegemony of UNIP over the labour movement. In practice, the one-party state's lack of control over Zambia's mono-economy made it increasingly difficult to make the concessions necessary to address the enduring expectations of mineworkers. The logic of the growing economic crisis and the threat to UNIP's political supremacy that flowed from it led the ruling party to assert its dominance in increasingly repressive and overtly undemocratic ways. This culminated in the conflict over decentralisation and to a breach in state-union relations that was only resolved in the multi-party movement of 1990-1991 (see Chapter Five).

There is little or no evidence that UNIP succeeded in gaining the loyalty of ordinary mineworkers. The decline in strikes was achieved primarily because of the explicit illegality of industrial action, not because mineworkers were convinced that their grievances were being addressed. Those strikes that did take place demonstrated the continuity of mineworkers' major grievances since Independence: racial and class conflict between supervisors and workers, expatriates and Zambians; and the unequal distribution of the wealth generated by their labour power. Whilst the state was able to limit industrial action at the point of production, these grievances were increasingly expressed in the mine townships, which fell outside effective state control. They were expressed by miners, but also by their wives, whose lack of representation paradoxically freed them to play a leading role in representing them.

As Nordlund argues, the detention of union leaders in 1981 had unintended consequences:

> Instead of discrediting the union, the Party had created heroes or martyrs in the eyes of substantial parts of the population. It was in retrospect then the starting point for the ZaCTU [ZCTU] as an unofficial opposition, with Chiluba as its leader and recognised as a political force by a large section of the population in general and in the Copperbelt Province in particular.[226]

Contrary to the state's claim that union leaders were responsible for working-class unrest, the events of 1981 also showed that mineworkers' resistance to UNIP policies was based not only on effective rank-and-file organisation in the workplace and the townships, but also on support for union leaders that was contingent on what those leaders did.

5

'THE HOUR HAS COME AT THE PIT', 1981 - 1991

Introduction

By 1981, Zambia's labour movement was widely identified as the *de facto* political opposition to UNIP rule.[1] During the next decade, the ZCTU leadership moved from cautious critic to active political opponent, ultimately playing a vital organisational and leadership role in the democratic movement of 1990. This led to the establishment of the MMD, which defeated UNIP in 1991 under the new multi-party system for which the ZCTU successfully campaigned. Researchers subsequently sought to explain the increasingly important political role of the labour movement during the 1980s.[2] Nordlund argued:

> … the single most important resource for having an impact on politics appears to be economic strength. The organisational capacity and position in the national economy of the union movement in Zambia made it possible to carry out successful bargaining with the ruling elite, and on both economic and political issues.[3]

Both Nordlund and Rakner identified the union movement's capacity to maintain its autonomy from UNIP, which enabled it to play such a prominent role in the MMD. They also note the importance of membership pressure on union leaders in ensuring this autonomy. Neither, however, explains the ways in which the actions of local union activists and members influenced the stance of the ZCTU, nor their motivations for these actions.

Some analysts argued that the ZCTU offered a 'business-oriented' critique of UNIP's state-based economic model.[4] In fact, as this chapter will demonstrate, throughout the 1980s the ZCTU leadership consistently attacked UNIP's implementation of pro-business structural adjustment policies. The labour movement criticised theincreased role of the IFIs in the management of the Zambian economy and the close relationship between multinational capital and state-owned corporations. This analysis found widespread resonance amongst the urban Zambian population. Contrary to their misleading portrayal as UNIP's core constituency, urban Zambians were in the forefront of opposition to UNIP's attempts to implement adjustment programmes.[5] In 1986-87 and again in 1990, they rejected the implementation of such programmes through widespread riots and strikes that reversed Government policies and damaged UNIP's hegemony. Periodic attempts at the incorporation of organisations representing urban Zambians into UNIP structures did not reflect their status as part of the party's constituency, but rather demonstrated the opposite – that they were dangerous and effective opponents that needed to be silenced through incorporation. Mineworkers saw the appointment of MUZ leaders to UNIP's Central Committee in 1988 as a threat to union autonomy and effectiveness. Local MUZ leaders overcame state intimidation to oust these individuals and elect those whose autonomy from UNIP would, they believed, be more likely to ensure the realisation of their expectations.

Increased state economic participation in the form of ZCCM, established in 1982, led to the greater use of mine revenue for the enrichment of UNIP leaders. Widespread discontent with the apparent impact on their living conditions led most mineworkers to believe that the removal of a corrupt state from their industry would enable these conditions to be improved. Mineworkers were increasingly prepared to take direct action to address their grievances, if necessary in direct opposition to the MUZ national leadership. As the economic crisis worsened, as industrial action proved insufficient to prevent their declining living standards and as state intervention in their industry and union increased, mineworkers came to believe that their economic exploitation could only be addressed by political change. In the late 1980s, many local MUZ leaders organised underground political opposition to UNIP, providing a powerful base for ZCTU National Chairman Frederick Chiluba and assisting his election as MMD and Zambian President in 1991.

Mineworkers' local workplace and community organisation enabled an effective expression of their expectations and grievances. They elected local leaders who were able to resist state repression, overcome undemocratic MUZ structures in order to challenge and replace the union's leadership and provide a vital organisational base to the multi-party movement of 1990-91. They were not, however, able to articulate independently these interests at the time when the overlapping leaderships of the ZCTU and MMD followed international ideological fashion and donor pressure in accepting the policies of economic liberalisation that they had hitherto consistently rejected. In following their leaders into the MMD, mineworkers surrendered the union-based autonomy that had informed their resistance to unpopular UNIP policies since Independence.

Unions as opposition

By the 1980s, ZCTU affiliates had been significantly strengthened by an investment in organisation, staff training and resources, enabled by the more secure flow of income under the IRA. Congress organised a wide range of training courses for stewards, branch officials and leaders. Rather than creating the efficient and disciplined unions envisaged in the Act, improved union organisation enabled increasingly effective industrial action.[6] This more balanced and effective union movement also strengthened the authority of the ZCTU leadership, no longer dependent on MUZ resources.

ZCTU National Chairman Chiluba and Secretary General Zimba, consistently re-elected to the ZCTU leadership, widened their critique of UNIP's structural adjustment policies, unhindered by unconvincing declarations of loyalty to Party and Government. They were in the forefront of attacks on the disastrous foreign exchange auction introduced in 1985 (see below). Whilst UNIP's implementation of structural adjustment was certainly not voluntary (see Chapter Two), the logic of such policies nevertheless led to renewed attacks on industrial militancy. In October 1982, President Kaunda condemned illegal strikes that damaged the economy and scared away prospective foreign investors.[7]

The Prices and Incomes Commission established in 1978 failed to prevent the deterioration of real wages; Congress consistently demanded compensatory pay demands to match rising inflation.[8] Such demands put the labour movement in conflict with IFI requirements for the removal of price controls and the imposition of pay limits. In 1983, the ZCTU General Council declared, 'the control and management of this country's economy has been shifted on to foreign international financiers.'[9] The Bank of Zambia Governor advised unions to stay within a 10% limit, claiming that bargaining above this level would derail the IMF programme.[10] The ZCTU responded that it could not be held to an IMF agreement about which it was not consulted.[11] In 1984, Congress rejected the statutory 5% wage increase limit, pointing out that actual inflation was at least 25%, double the official rate.[12] Failed attempts to maintain wage limits were abandoned in favour of free collective bargaining.[13] This put the onus on unions to win wage increases and led to an upturn in industrial action. Congress demanded a K50 per month general increase for all low paid workers as compensation for price de-control and threatened a general strike if this was not met.[14] Zimba

> ... cited the Tunisian case where the workers and consumers took spontaneous action (demonstration) against bread price increases resulting in the sacking of the Prime Minister ... the possible actions at workers['] disposal during such situations were strike, go-slow or work-to-rule, demonstrations etc.[15]

The press portrayed the labour movement as reactionary and sectional, in conflict with UNIP's representation of national interests.[16] The ZCTU responded by criticising conspicuous consumption by the state elite.[17] Unions demanding the K50pm pay rise pointed to expenditure on luxury imported vehicles for celebrations of the 20th anniversary of Independence.[18] Chiluba criticised leaders who discussed socialism but failed to implement the Leadership Code.

Union control and participation

During the 1980s, UNIP used various methods to undermine the labour movement's role as unofficial political opposition. Its approach was however contradictory; at times it utilised incorporative tactics and at others made more direct attacks on the powerful centralised Congress organisation that was the legacy of the 1971 Act. In 1983, Kaunda directed all organisations affiliated to UNIP (including the ZCTU) to operate within the Party's rules.[19] The threat to turn the ZCTU into a 'mass organisation' of UNIP (akin to its Women's and Youth Leagues) was periodically made but never implemented.[20] Zimba nevertheless urged the MUZ

> ... to fight hard and maintain their independent status. He said moves were underway to change the law so that Trade Unions can be brought directly under the party and to make provisions for appointed leaders to take over the reigns [sic] of the Trade Unions. If this happened, workers rights would be threatened.[21]

MUZ's 1984 Conference resolved:

> … the essence of any socialist principles is to transfer power both political and industrial to the workers and peasants, but … the manner in which the Party and its Government wishes to introduce mass organisation in this country would merely transfer power from workers to management, and this leaving the Zambian worker at the mercy of Management … We therefore suspect that turning the labour movement into a mass organisation has been prompted by the desire of the Party and its Government to control trade unions and not anything else. The Conference therefore totally rejects turning the labour movement into a mass organisation…[22]

In 1984, when Chiluba and Kaunda shared a May Day platform, Chiluba praised Kaunda, but also said that UNIP should be less utopian and more practical and that the Party and its Government had a duty to deliver essential goods to the people.[23] The following May Day, all speeches by labour leaders were banned and the ZCTU in turn boycotted May Day events.[24]

Increasing levels of industrial action led to the introduction of Statutory Instrument No.6 in January 1985. This empowered the Minister of Labour to revoke the statutory deduction of member subscriptions to trade unions representing industries in which illegal strikes had taken place. The ZCTU protested, claiming the measure contradicted ILO resolutions.[25] When SI No.6 was implemented against the Zambia Union of Financial Institutions and Allied Workers' Union (ZUFIAW), workers in all financial institutions struck in protest. ZUFIAW National Chairman Benedict Chikoti insisted that the union had been restraining workers from strike action.[26] Zimba argued that unions should not be held responsible for strikes, because in twenty years there has 'never been a strike caused by a trade union'.[27] The measure did not achieve a downturn in strikes and business representatives argued that it hindered good industrial relations by weakening union leadership.[28]

Whilst UNIP sought to weaken the labour movement, the ZCTU also expressed contradictory positions towards political participation. Congress complained about its exclusion from official bodies, but also withdrew union leaders from some parastatal boards.[29] Congress sought to fulfil its participatory aspirations, but was frustrated by its inability to influence decision-making. This was reflected in its initial ambiguity regarding the appointment of labour representatives to the UNIP Central Committee in 1988 (see below).

Business links and economic liberalisation

In the mid 1980s, the ZCTU strengthened links with local business representatives, with which it cooperated to resist UNIP interference in workplace decision-making. In July 1985, a ZCTU-Zambia Federation of Employers (ZFE) seminar called for the abolition of Works Councils and Party Committees at places of work, because they were confusing and weakening the role of trade unions.[30] Later that year, the ZCTU and ZFE jointly protested against proposed amendments to the IRA that would make it compulsory for union officials to be UNIP members.[31] Such links provided a foundation for the subsequent cooperation between business and unions in the MMD. This did not mean that the ZCTU had changed its position regarding economic liberalisation. Indeed, during the implementation of the foreign exchange auction, local business and labour were frequently allied in criticising international capital and its cooperation with the state.

In 1988, Newstead Zimba criticised the Government's rapprochement with the IMF:

> The Workers and their leaders are not prepared to take bitter pills on the I.M.F. again. The conditions so offered made people's lives difficult to sustain. ZCTU will speak out even it means scrapping it … We should not sell our Independence to the International financial institutions.[32]

As late as 1989, the ZCTU was arguing that

> … the Party and its Government are going back to the IMF because they have failed to make the strategy of growth from our own resources a reality … the move is very retrogressive. It will undo even the little gains or stability that came with the New Economic Recovery Programme. It will make life unbearable due to the massive social dislocations it will engender.[33]

ZCCM and the MUZ

In February 1982, Timothy Walamba was elected National Chairman of the MUZ, defeating the incumbent David Mwila by 51 votes to 37.[34] Walamba's prominent role in 1981's strikes and campaigns (see Chapter Four) contrasted with a widespread belief that Mwila had been softened by his proximity to Government and Party leaders.[35] Miners expected

Walamba to demand improvements in wages and conditions more assertively.[36] Walamba however emphasised that he sought good relations with the mine companies and publicly declared himself a loyal UNIP member.[37] Nevertheless, rejecting allegations of foreign interference in the labour movement, Walamba argued, 'It is not easy to destabilise a country in which citizens work on full stomachs.'[38] He defended the right of unions to comment on issues beyond simple industrial relations:

> Comrade National Chairman lashed at people who accuse trade unions of poking their noses in matters which do not concern them … there was no matter which did not concern the labour movement…[39]

Whilst the 51% nationalisation of Zambia's mines in 1969 (see Chapter Three) had little direct impact on mineworkers, the merger of RCM and NCCM into ZCCM in 1982 was widely experienced as increasing political interference in their working lives. Three months after it was established, ZCCM announced new austerity measures: recruitment was restricted and capital expenditure (vital to the industry's international competitiveness) was deferred. The need for increased workplace discipline was emphasised; an appeal was made to MUZ to avoid wildcat strikes.[40] The IFIs once again emphasised the centrality of ZCCM to Zambia's future economic development and their prominent role was made clear to MUZ:

> the performance of the Mines in the next five years was of crucial importance both to the Company and the Nation as a whole … The World Bank had people going round our Divisions checking on how we were implementing the various measures we had pledged to implement. Any shortcomings were being reported back to the World Bank.[41]

MUZ leaders criticised the implementation of cost cutting; Walamba suggested that mineworkers were paying a disproportionate part of the cost.[42] Miners were increasingly dismissed for minor offences.[43] Respondents associated the harsher industrial climate with increased political interference. Harrison Nkonde argues that 'when it was changed to ZCCM, it was more or less run by the Government.'[44] George Lombe agrees:

> Government was there, taking away the money, never reinvesting. Meaning that it now became a threat to our jobs. Those of us who were able to see that, stood up and spoke … against it. We wanted money to be ploughed back in the industry so that more jobs could be generated.[45]

Declining pay and conditions

ZCCM's need to reduce costs prompted accelerated Zambianisation. The first Zambian General Managers were appointed in 1983 and by 1984 there were only 1,800 expatriates in the mining industry, compared to 7,800 at Independence.[46] MUZ members continued to complain about the realities of Zambianisation. R.E. Mapulanga, who had taken over a previously expatriate position, wrote to Luanshya branch: 'Zambianization is taken to mean inheritance of job title ONLY and not the money … What treachery!'[47] Clerks in Luanshya's Personnel Department complained:

> … we are being used like experimental tools. We are regarded to be essential workers but the Personnel Department has turned a blind eye on us … Zambianisation in Personnel Department hasn't done a thing for us.[48]

Complainants bemoaned the decline in their living standards:

> … our salaries are peanuts on which we cannot depend. We have to utilise our normal off-days and hours to go into the bush to cultivate and subsidise our meagre "salaries" … Workers are not machines and treating them in this way does not augur well for industrial peace and stability…[49]

The first pay agreement negotiated with ZCCM in 1983 remained within the national pay limit of 10% insisted upon by the IFIs.[50] Blame for the unpopular agreement was laid at UNIP's door by the MUZ leadership:

> Some people have been coming up to challenge MUZ as having been taken for a ride by ZCCM which is not true. The truth of the matter is that pressure is coming from outside and that the Party and its Government is in control of the assets in the Mining Industry…[51]

In 1985, MUZ demanded the K50pm increase put forward by the ZCTU (see above) to compensate for rising inflation.[52] ZCCM argued that 'huge salary increases (such as those proposed by the Union) only scared away investors and antagonised the Company's creditors. This in turn would jeopardise investments in new projects.'[53]

ZCCM cited declining productivity to justify pay restraint: in 1975, each employee produced 11.2 tons of copper per annum, but by 1984 this had fallen to 9.6 tons. MUZ however argued that 'the decline in productivity was not entirely the employees' fault … FOREX [foreign exchange], spares and other production inputs were either lacking or operating ineffectively…'[54] ZCCM nevertheless told MUZ:

> The Company was not in a position to grant large salary increases and however unpalatable this message had to be conveyed to employees. Leaders had to accept responsibility for carrying bad news when the situation so demanded and that position had now been reached.[55]

Agreement was eventually reached for an approximate K50pm increase.

Strikes and conflict in MUZ

Short localised strikes continued to arise from similar grievances as in the 1970s (see Chapter Four). The more repressive political context and the increasing inability of ZCCM to make concessions tended to make strike action more confrontational.[56] A younger generation of more educated branch leaders was increasingly prepared to publicly challenge management. Branch officials met regularly with local management and, taking on representation of grievances previously addressed by the abolished mine township councils (see Chapter Four), they extended local MUZ authority to community-based issues.[57] Alex Lungu, Nchanga Branch Chairman from 1982, sought to re-define the illegal strike:

> Chairman Comrade Lungu stressed that as in most cases it was the Management that provoked the situation … such a strike should not be termed as illegal by [sic – but] legal … people were saying a legal strike was that sunctioned [sic] by both the Union and the Management, "where on earth will you find such a strike?" the Chairman asked … a Union without the right to strike was meaningless for a strike was a workers' loudest language.[58]

Industrial conflict was inevitable in the view of local officials:

> ... you cannot be a friend to management, as a trade unionist. You cannot be a friend to Government, as a trade unionist. Because they will never authorise you to go on strike. The fight that the unions have is the fight against the Government. And the fight the Government is having on the union, is to make sure that they are not strengthened, they are suppressed.[59]

Local MUZ leaders were monitored by intelligence agents.[60] According to George Mukoma, Konkola Branch Chairman from 1990:

> ... when you stand up as a Union official...you'd be very much aware that you got ... the Government intelligence wing ... being a public meeting, even police officers were coming there ... So each time you tried to criticise a certain policy ... I was called on four occasions to the police station. To try to give out the information, where I got that information and so on. And my most [important] weapon was ... if miners hear that I'm at the police station they'll go on strike. So they ... released me.[61]

Whilst Walamba was prepared to criticise ZCCM and UNIP, he no longer advocated industrial action to address miners' problems. Increasing criticism of Walamba focussed on his lack of communication with members.[62] When Roy Yumbe accused him of siding with management, the entire Nkana branch leadership was removed.[63] The Industrial Relations Court criticised MUZ's limited democratic accountability, particularly its lack of valid rules for the election of officials.[64] Similar concerns were expressed at Nchanga Branch, where 'The shopstewards expressed doubt about the presence of democracy in the union ... They accused the [Supreme] Council of having disregarded members in decision making...'[65] Walamba was unfavourably contrasted to Chiluba's outspoken leadership and his articulation of mineworkers' grievances. In 1984, Walamba was isolated in the MUZ Central Executive over UNIP's threats to turn the labour movement into a mass organisation of the ruling party:

> The National Chairman ... cautioned that this issue was likely to culminate in industrial unrest and hence he had informed the ZCTU

> that the MUZ was not going to be in the fore front [sic] so that we are not accused by the authorities as being the agitators when the situation becomes tense like has been the case in the past. But the delegates thought that such a stand would be dangerous because it would erode away the spirit of solidarity and weaken the labour movement. The meeting resolved to go ahead with the meetings…[66]

In October 1984, Luanshya stewards passed a vote of no confidence in Walamba (Nkana and Nchanga branches passed similar motions).[67] Luanshya Chairman Robert Kapenda confronted Walamba over President Kaunda's claim that pay negotiations had been concluded:

> … you, National Chairman, you are being congratulated … You accepted the handshake. Meaning you have sold the Union down the drain, I told him in the face … You are the mole in our midst … My loyalty was not to the National Executive, my loyalty was to the membership that I led in Luanshya…[68]

Kaunda was aware of how his praise of Walamba would be perceived:

> This cooperation does not mean weakness by MUZ, as some disgruntled people might say, but it is merely a reflection of the able and mature leadership of the union … I know many stupid people will say Walamba is a stooge because he is even praised by the President. I urge you comrade Walamba not to listen to them.[69]

Copperbelt MCC Alex Shapi publicly warned 'rebels' in the labour movement that the Party and its Government would intervene to stop attacks on Walamba.[70] This only confirmed the negative perspective of many MUZ activists:

> The relationship was very cordial between the Government … and the Union … [Walamba] was very much in favour … of the Government … that's what made him very unpopular … the general membership were suspicious of their leaders … So consequently they didn't even want to follow what they were saying … They were upholding what was in the Government, [rather] than what was required from the membership.[71]

In March 1985, Kapenda and his Branch Secretary Tony Malunga brought their complaints to the Supreme Council. They cited Walamba's 'failure' in negotiations, and declared that 'The National Chairman was praised by … Dr Kaunda which meant he was a sell-out.'[72] The Supreme Council voted for their expulsion, although they were subsequently reinstated by the Industrial Relations Court.[73]

Walamba was nevertheless re-elected at the 1986 Conference, defeating the challenge of his Deputy Charles Mbao. Interviewees report the buying of delegates' votes with Government funds provided for this purpose.[74] Boniface Kumwenda:

> Nkana branch … we have six executive members … these are the people who used to go for elections … And it's very easy for those people to be convinced. And Walamba, he has support from Government … Because someone in Government wants me to stay in power … It's very easy for me to get a loan from the Government … I don't think they were paying back those loans.[75]

The Mukuba Pension Scheme

A new mining industry pension scheme, 'Mukuba', was introduced in 1982, to which miners had to contribute 5% of their income. Following initial protests by Walamba, ZCCM agreed to contribute a matching 10.7%.[76] Mukuba funds were invested in parastatals and the scheme provided loans to MUZ-owned businesses. The perception spread that Mukuba funds were corruptly benefiting Government and MUZ officials.[77] Mineworkers feared that the misuse of funds would prevent them receiving their expected pension.[78] George Mukoma recalls, 'we [had] been cheated by Government … the word that filtered through … that Government was benefiting better than the employees themselves…'[79] As the first miners retired with small Mukuba pensions, demands were made that the scheme should be voluntary.[80] Nchanga Branch Chairman Alex Lungu attacked Mukuba at a public meeting in February 1985:

> … the Pension Scheme … did not benefit a miner at all … a Zambian life span had been calculated at 60 years, so if a miner should get his money after 55 years then hi [sic] was left with only 5 years to enjoy that money that he had saved for so many years…[81]

Walamba publicly defended Mukuba and a MUZ review resolved to maintain the scheme's compulsory nature.[82] Head Office refused to accept petitions for the scheme's revision from four branches.[83] On 1 June 1985, Chambishi miners, led by Branch Chairman Bonard Sekwila, went on strike, calling for Mukuba to be abolished.[84] The strike, which spread to Chibuluma and Mufulira, was directed against the union leadership rather than ZCCM and was particularly violent. Strikers burned the Mufulira branch office and looted the MUZ farm funded by Mukuba. Several people were injured. Police, having received death threats from strikers, were afraid to intervene.[85] In Mufulira, branch officials were summarily dealt with. Mary Palesi remembers that 'we were trying to talk to them but they couldn't listen unto us … most of our officials were really beaten … They were even burning the[ir] houses.'[86] Harrison Nkonde recalls how the strike was spread from Chambishi:

> In the plant and works area, instructions were issued … that this particular bus should go to Chingola. This … bus should go to Nkana. Reach the plant area. If they fail to open, stone them. Enter, go to the shafts. Engineering department. And ensure that confusion is all over.[87]

Worried management and union officials discussed the strike:

> … some employees who worked at Chambishi … forced their colleagues to take strike action … some buses were then commandeered from Chambishi mine and driven to Chibuluma, where on arrival they forcibly entered the mine area and forced everyone to abandon their work … [company] representatives went on to say that in view of the circumstances, a press statement by the Union was required which would strongly urge striking workers to return to work and to adhere to the laid down grievance procedure. As matters stood it was in the interests of both the Company and the Union (and the country as a whole) to maintain industrial peace. The Union representatives stated that they had been taken unawares…[88]

MUZ, appealing for a return to work, conceded that the pension scheme could be re-examined. Most strikers returned to work by 5 June. The strike cost ZCCM K16m in lost production. MUZ confirmed that, 'in order not to antagonise its membership, the Supreme Council has decided

… to recommend to both the Board of Mukuba Pension Scheme Trustee and to ZCCM that the Scheme be turned into a voluntary one…'[89] ZCCM agreed; although the strike had been violent, it had achieved its aim, again reinforcing mineworkers' perception that such action was necessary to have their grievances addressed. In the words of Kramer Nyondo, 'the Mukuba pensions board refused to refund the money. Until people went on strike. That's when the money was refunded to the miners.'[90]

Around 80% of ZCCM's workforce now withdrew from Mukuba. 845 strikers were dismissed, mostly from Chambishi.[91] The branch protested the dismissals to MUZ head office:

> Mr Bonard Sekwila was fired … Then we said how come that our strong Chairman has been fired from the Company? We realised that it was Mr Walamba's dealings. So we decided to walk to Katilungu House, to go and attack Mr Walamba … from Chambishi, on foot … we were turned back by the police … 'this particular procession is illegal' … When we came back … we passed a vote of no confidence in Mr Walamba and his entire Executive that they were all thieves.[92]

Following the strike, the Government used Statutory Instrument No.6 against MUZ, ordering ZCCM to halt deductions of subscriptions.[93] Exemplifying the contradictions in state labour policy, this was an attempt to demand from the union that it control its members despite its evident recent failure to do so. ZCCM management privately admitted the limits of legal restrictions on industrial activity:

> Despite the fact that virtually all employees in the Mining Industry are categorised as "Essential Workers" under Section 3 of the Industrial Relations Act, this has had very little effect, in our view, in curbing illegal strikes. Since 1981, for example, we have experienced unconstitutional strikes at Industry and Division levels to the extent to where one questions the efficacy of strike control through an Act of Parliament…[94]

For the union leadership, the resultant loss of funds was potentially disastrous. However, it also provided Walamba with a way to reconnect to an alienated membership:

> Authorities are all out to weaken the Labour Movement in this country. This can be evidenced by the introduction of Instrument No. 6 of 1985. Workers should rally behind their Unions for once the Unions are destroyed, their only safety valve in industry would have been destroyed.[95]

Walamba's unpopularity did not lead mineworkers to reject MUZ as an institution. Following campaigning by national and local leaders, 40,000 of MUZ's 50,000 members had completed voluntary check-off forms by September. SI No.6 did not permanently harm MUZ and President Kaunda waived its implementation in 1986.[96]

Resistance to structural adjustment, 1986 - 1987

Following the introduction of the foreign exchange auction system (see Chapter Two), the ZCTU's criticism of UNIP adopted an increasing class basis. In November 1985, Zimba delivered a wide-ranging critique of Government policies:

> Across-the-board prices of everything and anything are increasing every day … workers' real earnings are falling every day in direct proportion with the price increases … More realistic salary and wage increases will be demanded once the full picture of the auction nightmare has unfolded.[97]

The following year, Chiluba told the General Council:

> It is very clear now that there exist distinct classes in Zambia with identifiable interests and opinions which cannot be reconciled … People must have a moral, political and cultural right and duty to monitor the performance of their Government in all areas, criticize and provide the necessary check and balances … If one therefore wants to understand why the costs of living is rising? He has to trace the reasons to either political or economic policies and measures … This is why we are said to be behaving like an opposition party…[98]

The ZCTU consistently attacked the auction system.[99] In October, Chiluba condemned the IMF for putting African governments on collision course with their peoples and warned against the proposed removal of mealie meal subsidies.[100] Kaunda bemoaned the political influence of donors, but saw no alternative to adjustment policies. [101] Resistance came instead from the urban population, particularly on the Copperbelt. On 4 December 1986, following the doubling of the price of 'breakfast' mealie meal consumed by most urban Zambians, panic buying led to shortages. This sparked widespread looting and rioting in Copperbelt towns.[102] ZCTU General Secretary Zimba now advised the Government to review its links with the IMF.[103]

Following the introduction of the foreign exchange auction, ZCCM had awarded a unilateral 20% increase to its employees to offset price increases, doing the same in November 1986.[104] These were not, however, sufficient to prevent mineworkers participating in the food riots. Arguing that unionised workers represented a 'privileged' section of the working-class, Simutanyi claimed that ZCCM's maize meal subsidies meant that 'trade unions found it difficult to urge workers to resist maize meal prices, when the most militant of them did not feel the impact of the increase in maize meal prices.'[105] The food riots were not of course organised by the unions, but were a spontaneous local protest. Whilst mineworkers did not participate in these actions as a single and undifferentiated group, respondents report that, particularly in Kitwe's mine townships where the disturbances started, mineworkers rioted alongside family members and residents of informal settlements.[106] B J Ngulube recalls:

> ... there was some miners who were actually picked up at the end of the day. When Zambia Police came, the military carried [out] an inspection. [In] mine houses [in] Wusakile, especially the houses which are near the plant ... where the shops were actually destroyed. This is the area where goods were looted ... the Government found some items in these mine houses.[107]

ZCCM's sale of subsidised meal was limited to two 25kg bags per month. Whilst this might have been sufficient for the nuclear family imagined by the company, it was not enough to feed extended households of ten to fifteen residents. Miners' children and other kin were frustrated with their inability to find employment and were prominent in the rioting. Their concerns, however, were not opposed to those of the mineworkers, who were culturally obliged to feed their relatives. Whilst the riot has generally been subsequently understood as a spontaneous outburst of anger, for some it represented a form of politicised protest. Looting focussed on state stores and UNIP and Government offices were attacked, with youths directed by older men.[108] Sefelino Mumba, who took part in the riot, cites the burning of a poster of Kenneth Kaunda in arguing:

> … it was a matter of sending message to our brother [Kaunda], whatever you are doing we are, we don't want … We just threw there stones …'Kenneth Kaunda, what are you trying to do with us? Don't play with us. If we can, this building can fall. Please stop what you are doing.'[109]

A curfew was imposed in Kitwe and on 9 December the city centre was sealed off.[110] Fifteen people were killed by the police.[111] The rioting led to the restoration of food subsidies and was significant in ending Zambia's cooperation with the IMF.[112]

In the wake of the 1986 food riots, Chiluba claimed that further concessions to the IMF would make the rich richer and poor poorer. Zambia, he declared, was the only 'socialist' state implementing monetarist economic policies and strategies.[113]

The Medical Education Trust (MET)

As well as wage control, donor pressure on ZCCM to reduce costs led in 1987 to the introduction of fees in mine hospitals for township residents who were not members of mineworkers' immediate family. Mine hospitals and schools were hived off into a separate company, the Medical Education Trust. MUZ initially accepted MET, but it became increasingly unpopular amongst mineworkers and their families. As one nurse explains:

> … we have got extended families. We used to keep our brothers' children … And the money we were getting can't sustain when you have a patient [in the family]. And it became very hard for us. Even as nurses … [one] can see a patient coming, very sick, a non-miner, no mine number … when I came, shall pledge I'll preserve life. How do you preserve life, when you've got these acts again saying you should [only] treat somebody if that person pays consultation fee? [114]

In October 1989, miners' wives in Kitwe protested against the fees.[115] On one occasion, 500 women, some in overalls and helmets, besieged ZCCM's central office. As the *Times of Zambia* reported, 'The women, unmoved by the presence of armed police, chanted slogans and danced around the policemen, waving branches and calling for the abolition of MET.'[116] The wives threatened to picket to prevent miners going to work and, blaming MUZ for agreeing to MET, 'roughed up' union officials.[117] Ten women were arrested. Management representatives called on MUZ

> … to double their efforts in counselling their members to persuade their wives to stop the demonstrations … The Union should continue talking to their members through Shop Stewards to get the message through to housewives.[118]

Nkana Branch, however, voted to reject MET 'in its entirety'. Chairman Jackson Kaoma urged the Branch Executive Committee 'to bear in mind that protesting house wives and their dependants were an integral part of the mining community and therefore could not be isolated from mine employees.'[119]

MUZ subsequently demanded the abolition of MET, Walamba appealing to ZCCM not to allow the issue to fan industrial unrest.[120] Union criticisms expressed a rejection of MET's market orientation:

> Miners inspite [sic] of their poor health were being discharged from Hospitals earlier compared to those with money. Admissions also favoured those with ready cash … Prohibitive ambulance charges and medical fees had contributed to high mortality among extended family members of miners … MET's strong desire for money contradicted the Medical ethics of saving life first.[121]

ZCCM was angered at the reversal of MUZ's position, but sought to avoid alienating MUZ from its membership, since 'The Companies were aware that the industrial peace enjoyed only came about through a strong Union.'[122] MET was eventually abolished in 1991.

Underground political organisation

The riots of 1986 and the resultant reversal of Government policy revealed the one-party state's vulnerability to popular pressure. From this time, organised opposition to UNIP appears to have increased. Evidence of such activity, limited primarily to interviews, must be treated cautiously. However, it appears sufficiently extensive to conclude that in the late 1980s in the large mine towns, some mineworkers, particularly local MUZ officials, met secretly with other activists to discuss how to remove UNIP from power.[123] Enock Chansa explains that such meetings were held in peoples' homes, with activists pretending to be gathering for a meal.[124]

Evelyn Musonda, by then a Chief Shop Steward in Luanshya, was involved:

> There were a lot of meetings held at different points, even during the night. Although the police were monitoring each and every movement … They would think you're in a Union meeting but then you are planning your next move. Not only in MUZ offices, even in certain houses. Or, we would go like we are going to the fields and we would meet … we'll discuss A B C D, we meet and then we plan for the next day … we were in smaller numbers: eight, ten … we had informers … But we would just tell them, we know what you are doing and should you continue, we know if this thing leaks out we'll come and burn your house.[125]

Fidelis Mwamba, then a caretaker at Katilungu House, took part in secret meetings at Chiluba's house from around 1987, at which prominent trade unionists, intellectuals, and businessmen discussed the failings of the Government, and the need to form an opposition party.[126] Harrison Nkonde:

> … there were others among the workforce who were linked directly with … Mr Chiluba. Because in fact miners … were actually the less paid workers … a miner has to have something that can satisfy him. But those things were not forthcoming. So they realised that, should we go to them [miners], and use them, obviously we'll win … So these people were actually disseminating the information during night time … you would see probably around 13.00 hours, if it is Saturday the meeting will be held, Mr Chiluba and the others will move from all the way from Ndola, using the public transport … then he will just go straight to that particular house … Members will be getting [to] that particular house one by one … they shouldn't go in a group for fear of being identified … we realised that I think the best thing that will be done, to ensure that UNIP is brought down, is by uniting all the workers' representative organisations. Unions should be brought together. That's how we formed. 'Gentlemen the purpose of us meeting here, is to ensure that we come up with a way forward. To ensure that we get in power. How do we get in power?' Those were the same guys who had plans. Mr Chiluba, and the rest of the ZCTU.[127]

Japhet Sekwila links these initiatives to the subsequent creation of the MMD, and to the mineworkers' consciousness of their national importance:

> … the way that MMD started. It did not start as a … political party … this ZCTU President he knew, saying if there is a Province that I should earn support from, it's the Copperbelt. He knew that the Copperbelt is forcibly manned by the MUZ. So if he wins the support of the MUZ, he would win the support of the Province … he had [to] come straight to us … when you look at miner … it is the Copperbelt that feeds the nation…[128]

George Lombe was Mufulira Secretary of the Young Catholic Workers and a correspondent for its outspoken newspaper *Workers' Challenge*. Lombe was targeted by UNIP vigilantes for his actions.[129] Fear of intimidation checked the growth of such activities:

> … we were really being under state of emergency by then. And whoever talked against the Government, or UNIP … some of them were just missing mysteriously like that and you wouldn't know and because of that fear, things were just boiling in people's hearts…They were waiting … for a spark…[130]

For Joseph Mulenga, the defence of MUZ public meetings and the right to speak freely was intrinsically linked to political change:

> … the police had to say we have not given permission … but we had to dictate the situation and go up in the meeting … police had to come … and they lock you up. We had a spirit, actually that spirit was moving to do whatever we want … It is actually the Union which made the Zambian people to learn politics. That is the beginning that started to pave [the] way to democracy.[131]

Mulenga, a Nchanga shop steward, criticised Government policy in articles in the Catholic paper *Icengelo* in 1990. He was arrested by intelligence agents and tortured at the local police station.[132]

The increase in state intimidation was exemplified by the integration in 1986 of the mine police with the national police force. Now excluded from MUZ membership, mine police were given the 'full powers, privileges and protection accorded to regular Police'.[133] Their objectives included 'to preserve industrial peace in the mine area', and to 'detect and prevent crime, particularly industrial sabotage'.[134] Mine police thus dealt with both 'criminal' offences and the refusal of miners to go underground during industrial action.[135] Accusations of brutality (and complicity with intelligence agents) were now frequently made against the mine police.[136]

Members of the Central Committee

Walamba's limited criticism of Government and ZCCM policies ended in August 1988 when he and MUZ General Secretary John Sichone were appointed unexpectedly as members of UNIP's Central Committee (MCCs).[137] It is widely understood that Zimba and Chiluba had previously been offered and declined these positions.[138] Explaining the appointments, former Minister of Labour Frederick Hapunda told Nordlund:

> MUZ was, and still is, the union to captivate ... When you have a problem with the MUZ, it means that you have a problem with the whole country. The mining industry to this country is the whole lifeline ... Every effort was made to try to woo them to the Party, woo them to government ... There was a deliberate attempt, an obvious attempt, to alienate MUZ from the ZaCTU [ZCTU].[139]

Walamba claims the appointments were initially approved by the ZCTU:

> ... when we got that information that the President was going to consider appointing two from the labour movement, we called for emergency meeting ... each union had to send one delegate ... it was unanimously agreed that this is what we had always been looking for. And we were very happy to hear that Kaunda now finally is going to appoint two from the labour movement ... We went back to the Conference. Midnight, when President Kaunda came to announce the names ... Timothy Walamba ... and John Sichone ... to sit in the Central Committee. There was confusion amongst ourselves ... specifically, we wanted the ZCTU to sit as the umbrella of these unions. Nevertheless ... we unanimously agreed...
>
> What was going to be the reaction from the general membership? That was my worry ... the members outside will now be saying Mr Walamba, Mr Sichone, are sell outs ... I proposed to the ZCTU, I

said gentlemen, we have a problem. Because the miners are going to react … So we want you to come and address the Supreme Council, so that you can enlighten the general membership … So our first Supreme Council meeting was addressed by Newstead Zimba … Zimba explained to the Mineworkers' Union. And the Mineworkers' Union accepted…[140]

Zimba, and the Supreme Council, were in fact more ambiguous than Walamba recalls:

> … the Secretary General told the Council that … the labour movement as an entity should not be against the appointments but that it should be up to whoever might be appointed to make a decision … The decision was arrived at in order to forestall a confrontation.
>
> After this clarification a heated debate on the matter ensued with many councillors expressing fears about the appointments…it was agreed upon that Messrs Walamba and Sichone should seek for clarification from Freedom House [UNIP Headquarters] on what their role would be as Central Committee members … It is only then that the full implications of their appointment may be ascertained.[141]

Ordinary mineworkers unambiguously rejected the appointments:

> Walamba we said was a sell-out, he betrayed the miners … he was not even opposing the Government … he was there when they were formulating policies which were harsh on the miners.[142]

> … being a union leader and being in the Government, and [when] the Company belongs to the Government … I don't think you can fight for good conditions for the people, because … you discuss … the economy of the mine company, and they will tell you what to do…[143]

> … miners were saying … it meant they were bought by the Government. And the representation … will not be as effective as it was in the past … since he was given that appointment in a one-party participatory democracy … If Kaunda suggests to say he's not

> going to increase miners' salaries … there is no way, he's going to say, 'Kaunda, do increase'. No. That's why miners were very annoyed. He had sold them. You see? But we were going to get nothing.[144]

> … economically at that time, it was only the mines that used to produce, that used to sustain the well-being of the Government … if there was a work stoppage in the mine, just for five minutes, that's paralysis of the economy of the nation. So, because Mr Walamba was too vocal, the Government opted to silence him by offering that position…[145]

The litmus test would be the impact of the appointments on pay negotiations in 1989. The decontrol of prices that July led to an apparent promise of a compensatory 50% pay rise.[146] President Kaunda, however, argued that this depended on companies' ability to pay.[147] MUZ Nchanga Branch Chairman Albert Chali found himself caught between a national leadership prepared to accept a 12.5% pay increase from ZCCM and a membership unwilling to do the same. This led to violence at a subsequent public meeting:

> … the information had already leaked … I contacted the National Chairman … we have this problem in Chingola … to be on safe side, I wouldn't go ahead and address that public meeting. [He] said no no you have to. If you don't, we are going to discipline you … So I was entangled, you see? The members are saying we shall attack you … Towards the end of the meeting, when we're about to sing a National Anthem, they started throwing stones at us … the officer in charge [said] … the people are here, they've outnumbered our security men…we're going to be killed all of us. There were about 3,000 members … They [police] start now firing tear gas … they [miners] ran away towards this [MUZ] office, they started even looting. ZCBC was looted out, they came here, they found our office here, they burned everything … Until some of them were arrested, went to magistrate court. They gave me a chance to identify them. But you know as a leader, you cannot say this. So I had to defend them, to say no, they had a right to do anything.[148]

Two people were shot during the disturbance, and 31 were arrested.[149] It was reported at the time, and Chali confirms, that 'the [rioting] miners were demanding that MUZ National Chairman Comrade Timothy Walamba should choose to remain as their union chairman or as a Central Committee member.'[150] The 12.5% agreement confirmed miners' suspicions about the appointments, which were now publicly criticised by the ZCTU leadership.[151]

Rising strikes and Government response

As inflation continued to rise in the late 1980s, strikes occurred more frequently, particularly in the public sector. There was little sign that UNIP had any answer to such actions beyond their immediate repression.[152] Government papers provide an insight into its handling of the 1988 teachers' strike. Minister of Labour Unia Mwila was primarily concerned about the impact of the strike on Zambia's political situation:

> This year Zambia would have the Presidential and Parliamentary elections. It was necessary we have industrial peace so as to distract outsiders from undue activities which might disrupt political stability in Zambia.[153]

The Minister for Education criticised the failure of the police to take action against teachers:

> Security forces within Government were also blamed for laxity in not enforcing the law as they were aware that school teachers were on an illegal strike and more so, school teachers were declared as "necessary service" through the provisions of the Preservation of Public Security Act.[154]

The Inspector-General of Police explained the difficulties in ensuring the successful prosecution of teachers who had not been issued with Essential Service Certificates.[155] The criminalisation of industrial relations was largely ineffective. As Hamalengwa argues:

> What was also noticeable was the lack of resort to the IRA by the state regarding illegal and unconstitutional strikes. There were no prosecutions of illegal strikers. Neither were there any fines imposed against illegal strikers ... One of its most important functions - to prevent strikes - was aborted by the working class not obeying the dictates of the state and its laws.[156]

As well as prompting riots and a coup attempt (see Chapter Two), the removal of food subsidies in June 1990 also led to a wave of strikes by workers seeking a compensatory wage rise.[157] In August, a major strike paralysed all financial institutions, and was successful in achieving the reinstatement of sacked Zambia State Insurance Corporation workers.[158] In September and October, Zambia experienced an unprecedented strike wave affecting the motor trade, shoemaking, Zambia Breweries, bakeries and milling plants, postal services, colleges and schools, hotels, hospitals, sugar estates and textile factories. Coinciding with the period when the unregistered pro-democracy movement was testing its practical legality, there was a new confidence amongst workers that state authority could be effectively challenged. Despite the Minister of Labour's claims that the strike wave was organised by UNIP's opponents to create anarchy and usurp political power, little police action was taken against strikers.[159]

The 1990 MUZ Conference

Opposition to Walamba culminated in a challenge to his leadership at the MUZ Conference of March 1990. By this time, Chiluba had declared himself in favour of a multi-party system and the MMD was under construction. In the context of political upheaval, the position of MUZ was crucial:

> … everyone wanted to use Zambia Congress of Trade Unions, to achieve whatever they want to achieve … the most powerful affiliate of the Zambia Congress of Trade Unions are Mineworkers' Union of Zambia. And therefore, for one to win favours of ZCTU, one has to first win the favours of the Mineworkers' Union of Zambia.[160]

Some of those hoping to challenge Walamba were suspended from MUZ, or transferred to a different mine (thereby losing their union positions).[161] Walamba's was eventually opposed by Jonathan Simakuni, MUZ Deputy National Chairman. Walamba however believes Chiluba was behind him:

> I was called a stooge of Kaunda. So, Mr Chiluba had to build from there … he had gone through all the Branches of the Mineworkers' Union to campaign against me … he was sponsoring my Deputy, the late Simakuni, to take over from me. Because Simakuni was MMD.[162]

As Simakuni's supporters travelled to Livingstone, they were being watched. Evelyn Musonda:

> We had boarded buses … from Luanshya … We did not know that we were being followed by the … OP [Office of the President] people … Simakuni was likened to firewood, so we had this firewood on the top of the bus, all the way from the Copperbelt to Livingstone.[163]

In Livingstone, Musonda was questioned by police and her hotel room was ransacked.[164] George Mulenga was another Simakuni supporter:

> … we could have been killed there, because we were campaigning heavy for Simakuni. And then Government machinery came there, paying for Walamba. So it showed very clearly that they had an interest in the influence of the Union … these intelligence officers … all of them came supporting Timothy Walamba.[165]

The experience of Mary Palesi, Mufulira Branch Treasurer, reveals the extent of state mobilisation:

> … he brought in the intelligence … he sent his people to Livingstone. So those people were coming to our rooms. Walamba used to give them our room numbers … When they came to my room during the night, they knocked. I had to peep up at the window. My heart started pumping, what do I do? Now I'm caught! … What I did was, I had to leave my friend in that room, and jumped over [sic - out] the window…they were about four guys in plain clothes … two were holding their pistols. Which I saw. And the other two had AK47s … After some minutes I saw them moving to the next house … they were going into my friend's place. 'Where is Mary, we are looking for Mary? … we're looking for that young stupid girl! Who has made our big man so miserable.' After some couple of minutes, I saw them walk forward … They went to Mr Walamba's room. 'No we haven't seen this young girl'. [Walamba:] 'No, that is the very room where she is! Go back. Make sure that she's been caught.' Again they came back … The whole night, we couldn't even sleep. They were listening to us. But nobody was caught…[166]

The next day, Chiluba opened the Conference, repeating his call for a multi-party state and making an analogous attack on Walamba's position in the Central Committee.[167] Walamba claimed in contrast that

> … the appointment of the National Chairman and General Secretary into the Central Committee had enhanced the improvement of relations between the Party and its Government and the Labour movement on the other.[168]

Walamba lost to Simakuni by 73 votes to 50.[169] UNIP's attempt to control the labour movement through the manipulation of its leadership had once again been defeated.[170] Newspapers now claimed that Simakuni supported the ZCTU's call for a multi-party system.[171] Most interviewees insist that Simakuni was elected on the basis of his record in MUZ, but UNIP supporters argue he was already 'MMD'. For many in MUZ, there was an implicit association between effective trade unionism and opposition to UNIP. Police harassment of Simakuni supporters continued, as the new MUZ leadership complained to management:

> On return from Livingstone Conference, Wusakile Zambia Police Station delivered summon[s] notes to Nkana Branch MUZ Chairman and Secretary … it is alleged that Zambia Police accused the two MUZ officials [of] being anti-Party since they had voted out Mr Walamba who is a Party man … The General Secretary said that the MUZ was concerned about this incident as it could have provoked a disturbance…[172]

Nkana MUZ Branch Chairman Steven Mukuka had been one of the first local officials to favourably respond to Chiluba's call for multi-partyism. At Branch Executive Committee meetings, he described 1990 as the year of 'CHANGE' and 'ACTION'.[173] In June, Mukuka was briefly detained by Wusakile police, accused of inciting miners against UNIP.[174]

The MUZ and the MMD

Weeks after the MUZ Conference, the new leadership declared the union's support for the return of multi-party democracy. General Secretary Kossam Sheng'amo argued that the revamping of the economy required political change.[175] MUZ's call for political change was linked to mineworkers' enduring economic grievances. In October, when Simakuni offered his most detailed critique of Zambia's political economy, he did not reflect the MMD leadership's support for market economics:

> Is it not shameful that … ZCCM carries the unenviable stigma of offering one of the lowest and poorest conditions of service to its local employees, than any other mining Company in the world? It is mind boggling to note that ZCCM's top-heavy, politically inclined management at the helm of the country's economic main stay, could after 25 years of political independence, continue to maintain two separate and highly unequal conditions of service for expatriate and senior staff, on one hand and local staff on the other, as a vestige of the colonial industrial colourbar of the 1940's. In this unfolding political era, whose struggle and attainment this Union is proudly and closely identified with, we demand a fresh progressive approach to enterprise management and industrial relations on the part of the Company ZCCM … in which the satisfaction of workers' needs will match the pursuit of profits, at the expense of paternalistic affiliations to Party and Government institutions and structures.
>
> In a multinational and monopoly company dominated economy like ours, where even the state as a major employer tend[s] to be on the side of capital in labour matters … the role of MUZ and indeed all trade unions should have to be to safeguard all round democracy, individual human and civic liberties, social justice and the rule of law … the protection and promotion of workers gains in wages, conditions of service, living and housing standards, social security and welfare after retirement, against the extremes and excesses of *any future Government*…[my emphasis]
>
> In this hard struggle, we will have to bank more and more on the strength, unity and solidarity of all workers in alliance with revolutionary intellectuals to thwart the machinations of capital in all its manifestations, colours and shapes.[176]

Mineworkers' political and economic discontent, 1990

As with other workers, political upheaval provided mineworkers with space in which grievances could be more confidently expressed. In August 1990, an all-out strike took place in Luanshya over the non-payment of bonuses. Those seeking to work were assaulted and tear gas was used.[177] Annual pay negotiations provided an opportunity for mineworkers to express their political and economic expectations. In November, as negotiations were ongoing, workers at the Nchanga Open Pit struck.[178] Despite warnings from MUZ Head Office that the strike was illegal, the action continued into a third day. 2,000 miners demanded a 200% wage rise following the reduction of mealie meal subsidies in June, rejecting the 20% increase reportedly offered.[179] Following a return to work, 17 Nchanga 'ringleaders' were dismissed, amidst allegations that the police had beaten them to extract confessions.[180] Open Pit shop stewards leading the action were also local MMD leaders.[181] They included Sefelino Mumba, MMD District Youth Chairman:

> ... we went to general offices to deliver the message. So that we can give more chance for our leaders to aid discussion faster than what Management could accept. There was chants now going on...political slogans. 'The hour has come at Nchanga'.[182] ... So we went with placards ... even writing to [on] our heavy vehicles 'The hour has come' ... the Superintendent – Human Resources ... came to us and said, this kind of protest is political ... [They] ... said now [who] has planned this? ... Who's behind it? ... that's political ... it's MMD. So, I was mentioned...[183]

Mumba was one of 21 miners arrested:

> ... my name was called to give statement to panel of ten security officers ... another man told me that the case I committed was a serious one. That can result in me being imprisoned for life, or killed ... Questions from police officers: 'name the people you talked to.' I replied, 'workers'. Because I did not give the names ... I was removed [from] where I was seated, onto a wooden platform. And two policemen at the back handcuffed me, with a cloth at both eyes. Then took the long bar across my two legs. And suspended me, upside down. And then started beating me [on the back, with sticks]. Up for about 15 minutes.[184]

Mumba and others were charged with incitement to strike:

> The accused on 02/11/90 was found and seen addressing other employees at RTV Workshop and incited them to go on strike because the "Hour has come at the Pit" and that if they did not go on strike they would be given 20% increment.[185]

Mumba argued in his defence that the slogans were a political act and not a breach of ZCCM regulations.[186]

In the pay talks themselves, the new MUZ leadership broke with tradition by openly threatening industrial action. MUZ argued that 'Mine Employees were no exception to the current effects of inflation which had afflicted every Zambian and because of this, the situation in the Mine Townships was not normal.'[187] Criticising ZCCM's funding of political projects, the union 'likened the Companies to a parent who fed his children on a poor diet but assisted a neighbour in procuring delicacies', and warned that 'they were under extreme pressure from their members and could not therefore accept an offer which was going to infuriate their members.'[188] After the intervention of ZCCM Chairman & Chief Executive Francis Kaunda (no relation to President Kaunda), a breakthrough was achieved.[189] This agreement is today remembered as one of the greatest achieved by MUZ. President Kaunda subsequently claimed that advocates of the multi-party system had been calling on miners to sabotage the mines and told a UNIP rally in Chamboli, 'They [MMD] hoped the ongoing negotiations between the Mineworkers" Union of Zambia and ZCCM would reach a deadlock over salary increments, so that they would urge the miners to go on strike and damage equipment.'[190] Kaunda sought to portray the generous pay award as a way of averting such action. Mineworkers, however, attributed the pay rise not to the generosity of ZCCM or UNIP, but to the new MUZ leadership. The lowest paid miners received a K1,000 increase, an amount subsequently known to them as a 'Simakuni'.[191] This appeared to exemplify the better future that mineworkers could expect under the MMD:

> … people thought that … with the MMD pushing, already they are seeing changes, the Government is agreeing … that we should have annual increment … we were feeling that it's the … pushing of the MMD coming in. And people thought when MMD comes in then it will be easier for us … when we looked at our President [Simakuni] … we knew that he was MMD supporter. And when we looked at the Chairman General of ZCTU, as being an MMD supporter … this man, being a unionist, he must understand the plight of the workers.[192]

Simakuni's death

In January 1991, the MUZ Central Executive Committee reaffirmed its political position:

> There is no doubt that the greater majority of miners are MMD supporters and it is therefore only logical that the Union should give support to the MMD.[193]

Simakuni spoke at MMD rallies and was unambiguous in his support for the party.[194] In May, he suddenly died at the age of 54 after a short illness.[195] Chiluba's pledge that the MMD would try to establish the real cause of his death reinforced the widespread suspicion that he had been killed by the state. [196] This remains the overwhelming opinion of mineworkers today:

> … this thing is happening in the nation, where you see someone becomes a threat. The only option, you can't talk to him and he cannot compromise with you … is to eliminate him. His death is a political death. He was assassinated … politically killed, understand it.[197]

Mary Palesi:

> Simakuni as he was here today, miners were going to enjoy … Because he was a miner as well. He was a Winding Engine Driver. He had known all the problems that you was going through as a miner … He was just poisoned by the Government. 'Cos that man was strong … And he was in a opposition party, he was opposing the Government of the time.[198]

Miners and family members from across the Copperbelt attended Simakuni's funeral. Kramer Nyondo remembers that 'thousands of them attended the burial. All the shop stewards, from all the mines … all the MMD's supporters, they also came … It was a political rally; not like a funeral, like a burial.'[199] UNIP MCC Alexander Kamalondo was manhandled, and then chased away by mourners chanting '*Kaunda, kuya bebele, Kamalondo kufwa*' ('Kaunda must go, Kamalondo must die').[200] Many mineworkers believe Simakuni's early death robbed them of a leader they thought would meet some of their expectations for economic and social progress.[201] David Longwani's belief that, 'If only he had lived longer than that one year … things were going to change', is widely shared.[202] David Ndalama's description encapsulates the attributes mineworkers associated with Simakuni and all good union leaders, claiming that 'he had the people, their interests at his heart. People tried hard to bribe him … But [he] was able to shake them off. He was a man of character, with a lot of integrity.'[203]

The Industrial Relations Act (1990) and Pro MUZ

The Industrial Relations Act of 1990 removed the ZCTU's statutory powers established in the 1971 IRA, but maintained the Minister of Labour's powers to suspend or dissolve trade unions.[204] Unions seeking to support a political party were now required to establish a political fund, endorsed by two-thirds of members.[205] This was a tacit acknowledgement of the failure of the corporatist approach, but also punishment for the labour movement's support for the emergent political opposition.[206] The revised IRA had little practical impact, beyond providing the final impetus for the labour movement to formally declare its support for the MMD. As Akwetey argues, 'The demand for the restoration of multi-party politics … was directly related to the leadership's perception that the new IRA (1990) threatened the existence of the ZCTU and the organisational cohesion of the labour movement.'[207] In December 1990, the ZCTU General Council declared:

> The Act destroys all freedoms of expression … trade unions and their workers and all peace-loving citizens of Zambia are called upon to effectively and morally support the efforts being advanced by MMD which we believe will form the next democratic government.[208]

In January 1991, the amended Act was discussed by the MUZ Central Executive:

> … the Party and its Government had amended the Industrial Relations Act in which the major thrust of those changes was intended to punish the ZCTU and its affiliates for having spearheaded the reintroduction of Multi-party politics in Zambia … Unions in this country were going to face very serious setbacks in this situation … The meeting suggested that Head Office should reproduce and distribute to Branches copies of the amendments to the Act to enable them explain to the members properly.[209]

The amended IRA removed the provision for 'one union, one industry', and led to a breakaway from MUZ by artisan and graduate mineworkers. Following their strike in 1981 (see Chapter Four), an interim committee of representatives of these groups had been established in 1983.[210] Previous breakaways had been prevented by MUZ's monopoly of representation, but now the Professional Mineworkers' Union of Zambia (Pro MUZ) was established. Its interim chairman Percy Chanda claimed that professionals were not properly represented in MUZ, which only looked after unskilled miners.[211] Despite Chanda's contemporaneous and subsequent denials, it was widely believed that Pro MUZ was a politically inspired attempt to divide MUZ, so as to weaken its political impact. Although Pro MUZ represented the aspirations of skilled mineworkers, it only existed because of the new IRA, itself a political attack on the labour movement. Despite this, UNIP was also suspicious about organisations outside its control and failed to grant legal recognition to Pro MUZ:

> UNIP which was in power, was suspicious about our move. The MMD which was trying to come in power, was suspicious because of our moves. You would find UNIP is issuing a statement that those chaps are MMD. MMD is issuing a statement that those chaps are UNIP. So … we were actually targeted for victimisation.[212]

Whilst Chanda was followed by intelligence officers, MMD supporters in the mine townships who saw Pro MUZ as a UNIP front physically attacked its leaders.[213] The new union was effectively wound up in 1992.

The ZCTU and the MMD

Following Kaunda's concession of a referendum on multi-partyism in May 1990, the ZCTU declared it would campaign for the return of multi-party democracy. The following month, Congress dramatically reversed its economic policy, welcoming the proposed liberalisation of the economy and calling for complementary political restructuring.[214] Chiluba now maintained that, rather than economic inequality, it was Zambia's unhealthy political environment that was the cause of industrial unrest.[215] Whilst the ZCTU continued to claim that the IMF and World Bank had imposed an inappropriate economic liberalisation programme on UNIP, it was no longer opposed to such policies in principle.[216] In May 1991, ZCTU Head of Research Ages Mukupa approvingly cited the Scandinavian system in which, he claimed, capital accumulation was used to create social and material prosperity for all. Mukupa argued that in a genuine democracy, capitalist development could be pursued without worrying about creating extreme inequity.[217]

Industrial action continued to influence political rhetoric in the run-up to the 1991 elections. In May, Kaunda noted the rise in strikes since the return of the multi-party system.[218] In August, he claimed that public sector strikes on the Copperbelt were politically inspired. UNIP Secretary General Grey Zulu argued the ZCTU's organisational support for the MMD meant that 'workers' money is being used for pluralism instead of trade union affairs.'[219] Zimba, and MMD leaders, now argued that workers should 'exercise maximum restraint and persuade themselves to call off any industrial action they may be contemplating'.[220]

ZCTU leaders called for support for the MMD on the basis that it was 'a workers' baby'. As Akwetey states, 'The ZCTU placed its organisational network and facilities at the disposal of the MMD...'[221] Nordlund argues:

> ... the unions were seen as the organisational basis for the newly formed movement. By August 1990, ZaCTU [ZCTU] funding and organisational infrastructure were also made available for MMD use – instantly providing the MMD with a campaign vehicle that was widely superior to that of the ruling party.[222]

On the eve of the election, acting ZCTU Chairman Fackson Shamenda declared that the labour movement would criticise the MMD if it failed to deliver the goods.[223] After the MMD's election, Shamenda led a ZCTU march in support of the new Government. He called for workers to be disciplined and to be prepared to sacrifice, but warned that their patience would not last forever.[224]

MMD and the mineworkers' aspirations, 1991

Local MUZ officials played a leading role in establishing MMD organisation across the Copperbelt, its strongest area of support. In Konkola, although branch officials were active in the MMD, the party was prevented from using union facilities.[225] In Nkana, under Branch Chairman Steven Mukuka, the MMD met in the MUZ branch office. With the MMD's registration in January 1991, past and present local MUZ officials took up positions in the party. Abel Chisanga became MMD Chairman of T Section in Wusakile.[226] Harrison Nkonde was a Ward Vice Chairman in Mufulira.[227] Maxwell Kalesha-Chisoko chaired the 'Jonathan

Simakuni' MMD branch in Chingola.[228] Evelyn Musonda became an MMD Chair in Luanshya:

> We were free to organise because, if we saw a police vehicle pass through … we would ask our small children to just throw stones at the vehicle. They stopped coming to the compound because everybody was now just for MMD … There was nothing like victimisation … they feared that if they did this, then in the morning there'd be chaos. We just had to campaign freely.[229]

Albert Chali, MMD Chingola District Vice Secretary, claims that 'the Union made it possible for MMD Government to come to power. Because of the Union. The revolution started from the Union. Just like the revolution started before Independence, from the miners.'[230] Sam Kangwa argues that 'they were our own leaders … who shifted from union matters to political matters … people were even calling it … ZCTU-MUZ Party...'[231]

Whilst Timothy Walamba spoke at UNIP rallies, attacking those who had criticised his political appointment and who were now taking up a role in the MMD, his predecessor David Mwila became MMD District Chairman in Kitwe.[232] At vast rallies on the Copperbelt, the MMD explained their plans for the mines. In April 1991, Chiluba told 100,000 people in Chingola that, 'As for now the mines are centralised and the money they earn is being taken to Lusaka where a few individuals help themselves to it like kings. Such are some of the things we want stopped.'[233] The MMD attacked ZCCM inefficiency and the cost saving measures that had cut miners' jobs. The neglect of miners in the allocation of ZCCM's revenue was criticised; Chiluba claimed that the company's top executives were collectively paid more than the 55,000 miners combined, all of whose jobs would be safe under the MMD.[234] Such messages resonated with the experience and aspiration of mineworkers:

> … it was the state which was deciding how much they can put into the running of the mines … the whole money was going to the politicians…So that's why people … thought maybe if it is privatised, maybe there'll be some changes.[235]

> … they said that if we privatise the mining Company, it'll mean you miners will stay very well, your salaries … the conditions that you

> will be in will be very good, compared to the conditions that you have now ... they said ... they will create employment...[236]

> They said when MMD come into power, they will look at a miner as the baker of the national cake. Because during that period ZCCM, the national income used to come from [it], at least 80 to 90% foreign exchange came from the mines. But we never used to enjoy ... other people were enjoying it...[237]

UNIP, meanwhile, tried to use the privatisation issue to gain support from mineworkers. On the eve of the election, a UNIP advert declared:

> Miners be warned about the MMD
> Has the president of the MMD Mr Frederick Chiluba promised you improved conditions, but what do his policies say?
> i) He wants to split ZCCM so that each mine runs separately.
> ii) He wants to invite foreign investors for each mine.
> iii) He wants to give the private investors a free hand to improve efficiency.
>
> What does efficiency mean?
> Efficiency means more profits for the shareholders.
> How do shareholders achieve there profits?
> i) By declaring workers redundant.
> ii) By closing marginal mines such as Kabwe, Luanshya, Mufulira.
> iii) By reducing expenditure on workers' welfare in amenities such as football clubs, hospitals, townships, subsidies on housing, mealie meal, water, etc.
>
> Do you treasure your jobs?
> IF YOU DO, THEN VOTE FOR UNIP AND K.K. THEY HAVE ALWAYS LOOKED AFTER YOU AND WILL CONTINUE TO LOOK AFTER YOU.
> TO VOTE FOR MMD IS TO VOTE FOR UNEMPLOYMENT.
> ON 31ST OCTOBER 1991 VOTE UNIP AND KK [238]

Meanwhile, the MMD made other electoral promises that won mineworkers' support:

> … those [sub-standard] houses there, were going to be demolished. And they would build new houses, and a big market is going to be built there … And then they said a clinic was going to be built, [and] a bus station … they said, we are going to form a cooperative, which will help those people who had no jobs.[239]

The MMD promised that transport services would be improved, shortages of essential commodities would end and (despite assuring donors that subsidies would be removed) that the price of mealie meal would come down.[240]

At the same time, miners and other workers were marginalised in elections for senior MMD positions on the Copperbelt. As Bratton described, MMD leaders were selected for the Province's expected safe seats and money became increasingly necessary to secure local posts.[241] MUZ shop steward George Lombe, interim MMD District Youth Chairman in Mufulira, was defeated for the permanent position by a candidate sufficiently wealthy to fund an election campaign.[242] Only national MUZ and ZCTU leaders were selected as Parliamentary candidates. In Wusakile, for example, ZCTU Deputy General Secretary Chitalu Sampa was selected ahead of MUZ Branch Chairman Steven Mukuka.

Pay negotiations, 1991

As the election approached, Simakuni's replacement, Francis Kunda, criticised ZCCM during pay negotiations:

> … it had now become customary to receive a lecture at the beginning of each negotiation on the state of the Company, national economy and patriotism … while a miner was being called upon to sacrifice, others in the economy did not appreciate this and were having it easy … The Companies had recently bought vehicles for other people, put up a commuter railway line in Lusaka, built a hospital somewhere else. The above could not be done by a Company in a financial crisis…[243]

MUZ negotiators expressed optimism about the future:

> … the future Government is going to liberalise the economy and privatise the mines which will ensure efficiency, accountability and productivity. They hoped that ZCCM would behave better than now

> when they are engaged in all sorts of activities and seemed to be more concerned about running the Government.[244]

Following the election, the union expressed anger at ZCCM's refusal to increase its pay offer. However, MUZ was now more willing to reach agreement:

> ZCCM was just like the old UNIP Government which had lost touch with reality and failed to read the mood of the people...The new culture in the Third Republic called for an equitable distribution of resources ... It was probably that the Company wanted to draw the Union into some desperate action so that they could embarrass the new Government. They resolved not to do anything that would embarrass the new Government and pledged to continue negotiations until they reached an amicable settlement.[245]

Following a pay agreement, Kunda repeated the calls of his predecessors for an end to disparities in local and expatriate conditions of service, demanding that, 'The continued policy of maintaining two separate and highly unequal pay and other conditions of service between expatriate and local personnel ... should be done away with as soon as possible.'[246] Kunda outlined his hopes for a new era of industrial relations:

> As equal partners in the industrial relations arena, we demand not just a say but that our demands have to be accorded due understanding and respect. The picture of ZCCM as a top-heavy, almost self-barricaded management prone to paternalistic political affiliations and an over-ze[a]lous propensity to take on and over-burden itself with traditional Government responsibilities and several uneconomical ventures at the expense of lowly paid and overworked employees should be buried six feet deep with the discarded Second Republic-One Party misrule...[247]

Conclusion

This chapter has shown that, in a context of economic decline and increased state intervention in the mine economy, Zambia's mineworkers and their families were able to use industrial and community-based direct action to challenge their declining living standards and to resist attempts by UNIP to undermine and/or gain control of their union. In so doing, they expressed their economic and political discontents with the distribution of wealth and power in Zambian society and the consumption of the results of their labour by what they saw as a corrupt state-based elite. Despite a conspicuous lack of genuine democracy in MUZ, mineworkers were able to remove a national leadership that had, in their view, surrendered its independence by accepting positions in UNIP's Central Committee.

In 1991, most mineworkers expected that the new MMD Government they had helped bring to power would begin to address the radical decline in wages and living conditions they had experienced during 1980s. This was based in part on the prominent role that the labour movement as a whole had played in the MMD, both as pressure group and political party, symbolised by the leadership of Frederick Chiluba. It was reinforced by their participation in illegal underground political opposition, and subsequent MMD organisation on the Copperbelt. This participation, however, was not reflected in the effective representation of mineworkers' interests in the MMD. The ZCTU sought no direct influence over MMD policy making and had no explicit ideological basis or programme of demands to inform its relationship with the new Government. Their experience of, and discontent with, state ownership and administration of the economy and the relationship with both international capital and international finance institutions that this entailed, led them to accept MMD pledges that the removal of state intervention would improve salaries and living conditions, for themselves and other Zambians. This was to have significant consequences following Zambia's return to multi-party democracy.

6

'TO DIE A LITTLE': POLITICAL AND ECONOMIC LIBERALISATION, 1991 - 2005

Introduction

The expectations of mineworkers that the MMD Government that came to power in 1991 would address their enduring grievances were largely dashed. Mineworkers (and most Zambians) experienced severe declines in their living standards. More than half of the 56,500 workers employed by ZCCM in 1992 have been retrenched. Those still employed in the now privatised mine industry have seen drastic reductions in the value of their salaries and the loss of fringe benefits provided by ZCCM and its predecessors.

Whilst mineworkers have responded with periodic protests and occasional industrial action, they have not offered effective resistance to MMD policies. Neither mineworkers, nor the MUZ, have played a significant role in organised politics within the multi-party system they helped to bring about. This chapter seeks to explain why mineworkers saw few of their aspirations realised. It examines MMD governance in general and its economic liberalisation policies in particular. It explores the reactions of the labour movement to MMD policies, from initial support, through divided reaction, to a period of recent resistance.

The implementation of mine privatisation is explored, highlighting the corrupt nature of the process. The relationship between the Government, international mine capital and the IFIs, and the ways in which this shaped mine privatisation, is addressed, as is the marginalisation of mineworkers from this process. Finally, the chapter explores mineworkers' current perspectives on their participation in the events of 1990-91 and their attitudes to wider questions regarding the role of trade unions in political change.

MMD economic policy

When it came to power in October 1991, the MMD inherited an economy in near collapse. Zambia's dependency on international donors left it with little choice but to implement further liberalisation of the economy. Its programme was, however, one of the most radical structural adjustment programmes ever implemented in sub-Saharan Africa. All consumer subsidies were removed by 1992. Tariffs and other barriers to imports were removed. A decade-long programme to privatise 250 state-owned companies was carried out. State support to agricultural production was substantially reduced. The result has been a significant contraction of the economy, and a dramatic increase in poverty. As Rakner states:

> After almost a decade of uninterrupted policy reforms, the record in terms of economic growth, employment creation, investments and poverty reduction in Zambia is weak … the Zambian economy has shrunk and is now smaller than it was in 1991.[1]

Formal sector employment now accounts for just 11% of the labour force.[2] Social indicators drastically declined under the MMD. Primary school attendance had fallen to 68% by 2001.[3] Education spending as a percentage of Gross National Product fell from 4.5% in 1991 to 2.9% in 1994, although it has subsequently risen marginally. Health expenditure fell in proportion with overall cuts in Government spending.[4] User fees were introduced at all levels. Infant and child mortality has increased with the introduction of these policies and access to clean water and sanitation services has declined. Zambia is the only country with a lower Human Development Index in 1997 than in 1975.[5]

None of these measures significantly improved the competitiveness of the Zambian economy. From 1992 to 1998, Gross Domestic Product (GDP) declined by an average of 0.2% pa.[6] The closure of uncompetitive industries was not matched, as was forecast, by a revitalisation of the agricultural sector. The removal of agricultural subsidies in the early 1990s led to a drastic decline in maize production, which was only reversed by the re-establishment of limited state subsidies. Indeed, business interests that initially supported economic liberalisation became rapidly disillusioned. In 1993, farmers' organisations complained that agricultural liberalisation had harmed local production. As Rakner describes, 'When the business associations accused the structural adjustment programme of having butchered Zambian industry, the government was arguably left with a rather limited support base.'[7]

Structural factors continued to deter productive investment. These included high inflation and interest rates, Zambia's geographical isolation from major international markets and poor transport infrastructure. Most Zambians are simply too poor (80% live in absolute poverty) to support a functioning market in basic goods that would attract potential investors. Whilst some commercial farmers have benefited from the open market in maize and tobacco, the MMD's liberal policy on profit repatriation has prevented the benefits of these investments from spreading to the wider economy. Many privatised companies were asset-stripped and closed. The need to diversify the economy away from dependence on copper is, if anything, more urgent; Zambia's ability to do so is, as ever, in doubt. National decline remains as closely tied to the international copper price, as to the actions of the Government. The Government is even more dependent on external assistance than its predecessors; 53% of government expenditure is funded by foreign aid.[8] Despite the evident failure of economic liberalisation and privatisation policies, donor pressure has ensured their continued implementation.

Zambian businessmen and prominent politicians, many of whom accumulated their initial wealth through state capitalism under UNIP, have used the privatisation process to acquire smaller parastatals in dubious circumstances. For example, the $14.8m paid for the Zambia Sugar Company has not been accounted for.[9] Loopholes enabled a minority of Zambia Privatisation Agency (ZPA) Board representatives to implement decisions, without the participation of other board members.[10] By February 1998, 128 of the 188 privatised companies had been sold to Zambians. However, the 70 companies sold to foreigners were worth eight times as much.[11] Foreign investors were the primary winners from Zambian economic liberalisation. The negative impact of liberalisation policies has, in recent years, stimulated a popular movement against the privatisation of the few remaining state-owned companies (see below).

Politics in the Third Republic

The overwhelming victory of the MMD in 1991 initially undermined the potential for competitive multi-party democracy. The MMD's uneasy coalition of interests, the unity of which was vital to ending UNIP rule, rapidly began to fracture. Four cabinet ministers were dismissed in April 1993, strengthening the position of Chiluba and his allies. The MMD reneged on its electoral promise to reduce Presidential powers.[12] Chiluba's appointment of District Administrators replicated UNIP's system of District Governors in the central control and distribution of power. The loyalty of most MMD MPs was assured by their appointment to an increased number of Ministerial positions.

As disillusionment with the MMD's policies grew, the prospect loomed of electoral defeat in 1996. Zambia's Constitution was amended that year to disqualify Presidential candidates whose parents were born outside Zambia, a measure designed to exclude Kenneth Kaunda (still the leader of UNIP). The MMD won the 1996 elections in a poll boycotted by UNIP and marked by vote buying and the intimidation of opposition candidates. In 1997, a coup plot was used to justify the detention of Kaunda and other political opponents. Accusations of state-linked corruption increased during Chiluba's second term in office. Whilst this generated increasing popular discontent, opposition parties failed to offer significant policy alternatives. Rakner convincingly argues that the dominance of external donors in policy-making, via a series of economic liberalisation programmes, prevented domestic politicians from effectively presenting alternative policies.[13]

Multi-party democracy, however, provided space within which other organisations were able to grow. New independent media challenged corruption and criticised Government policy. When President Chiluba threatened not to stand down after ten years as President in 2000-01, civil society organisations, supported by popular demonstrations, successfully campaigned against his attempt to stand for an unconstitutional third term in office. This was, for some, a partial and welcome return of the 'spirit of 1991'.[14] Chiluba was instead forced to name former Vice President Levy Mwanawasa as the MMD's Presidential candidate. New opposition parties now emerged, most led by former MMD politicians who made serious allegations of corruption against Chiluba. However, despite their lack of policy differences, they failed to unite against the MMD. Mwanawasa won the Presidency, despite receiving only 29% of the vote, in a poll regarded by international observers as rigged.[15] The many opposition parties newly represented in Parliament did not provide a practical challenge to the MMD. The defection of opposition MPs and the utilisation of state resources to win by-elections subsequently gave the MMD the Parliamentary majority it failed to secure in 2001.

Mwanawasa rode the wave of popular desire to see those responsible for corruption brought to justice. In July 2002, he successfully appealed to Parliament to remove Chiluba's immunity from prosecution, a decision confirmed by the Supreme Court. In August 2003, Chiluba was charged with 96 counts of theft of $40m.[16] This was an unprecedented event in sub-Saharan Africa and widely regarded as a significant blow for democratic accountability. The legal process did not, however, expose the full extent of corruption under the MMD. The charges against Chiluba ignored the major corruption scandals of his administration, which raise wider questions about the complex relationship between the Zambian state, national and international capital. Two years after Chiluba was charged, the process remained bogged down in legal procedure. In September 2005, the Government invited a British court to sit in Zambia to hear charges against Chiluba, an unorthodox procedure which, it was hoped, would speed up the legal process.

Evidence of corruption has necessarily forced MUZ activists to reassess their view of Chiluba. Some believe that he successfully hid his inherently corrupt nature during his leadership of the ZCTU.[17] Others specifically note his betrayal of his former trade union colleagues:

> He killed all the unionist[s]. Those people ... they had no power ... no teeth. He forgot that he was once a unionist. He's a person who fought the Kaunda regime. How should he forget the unionist? He ... got all the power from the union leaders.[18]

Borries Ntankula locates this betrayal in Chiluba's links to capital:

> ... we didn't think that we are voting for a puppet. A few individuals today have benefited from the Zambian economy, but at the expense of large masses which are ... part and parcel of the economy of this nation. A miner today ... he has to go for work on an empty belly. School-going children go for schools, to attend lessons on empty bellies ... that's the kind of situation that MMD under Chiluba subjected every Zambian ... he was, and is still, a capitalist, who has amassed wealth, not only by himself, by a few individuals who were so close to him...[19]

The MMD and the ZCTU

In late 1991, President Chiluba wrote to the general secretaries of all Zambian unions, requesting their assistance in addressing Zambia's economic problems in stark terms:

> It is therefore my sincere hope that you will all summon courage as leaders of our people to convince the workers to die a little so that prosperity is not overburdened with a crippling debt and an economy shattered beyond redemption.[20]

In the early 1990s, the ZCTU leadership was generally supportive of the party it had helped bring to power. ZCTU President Fackson Shamenda confined his criticism to specific aspects of structural adjustment policies, meeting with the IFIs to request their more gradual implementation. Accepting that liberalisation was inevitable, Shamenda frequently demanded that the labour movement be consulted over its implementation.[21] However, the MMD generally excluded trade unions from its formal decision-making processes. This has been presented as a major shift from UNIP's corporatist approach.[22] In practice, the labour movement's *practical* influence over policy was as limited as under UNIP. It is notable, however, that the leadership of the MUZ continued to have direct access to State House.

Amendments to the IRA in 1993 and 1997 continued the one-sided liberalisation of labour legislation begun by UNIP in 1990 (see Chapter Five). Although the ZCTU lost its legal monopoly over union affiliation, the Commissioner of Labour retained substantial powers over union registration. The wide definition of 'essential services' continued to make most strikes illegal. The police retained powers to arrest those encouraging workers to strike.[23] Privatisation and civil service cuts led to vast job losses. Peaking at 358,000 in 1990, union membership declined to 245,500 in 1998, and continues to fall.[24] Having placed its hopes in the MMD, the labour movement lacked any ideological base from which to critique its policies.[25] This culminated in a split in the ZCTU in 1994. Four major unions, including MUZ, left Congress, partly because they believed there was no alternative to privatisation (MUZ rejoined in 1999).

In the context of increasing popular disillusionment with the MMD, the union movement has, since 2001, regained some of its former influence. The ZCTU developed links with non-governmental organisations and sought to provide support to retrenched workers. At the same time, an increasing number of public sector strikes challenged deteriorating pay and conditions and the non-payment of terminal benefits. The ZCTU became increasingly critical of structural adjustment policies, producing an important survey of their impact in Zambia in 2001.[26] In the 2001 election, both the ZCTU and MUZ argued for the MMD to be voted out of office, but did not endorse a particular opposition party. ZCTU President Leonard Hikaumba explains:

> I think the first few years of MMD in power, you know we had given them the benefit of the doubt. Most of the people in Government came from our movement, including the President.

> They had made an appeal in 1991 … the question was asked 'Are you ready to sacrifice?', we said 'yes' … And that sacrifice, in order to give chance to the Government, was also perceived as a weakness on our side, because even on things that we could have fought against in the previous administration, we allowed them to go … [Now] … we said, 'Look, for how long are we going to sacrifice?'[27]

In December 2002, trade unionists led demonstrations against the privatisation of the Zambia National Commercial Bank (ZNCB), and other state-owned companies. In response, the National Assembly voted against the sell-off of ZNCB, reversing government policy. Mwanawasa declared that he was personally opposed to further privatisations.[28] In March 2003, however, Government ministers announced the 'commercialisation' of ZNCB: *de facto* control of the bank would be transferred to a private foreign bank.[29] The Government's *volte face* followed the IMF's public warning that retaining the ZNCB and the other companies in state ownership would breach Zambia's Poverty Reduction Strategy Programme (PRSP), threatening the substantial debt relief possible under the PRSP.[30]

In February 2005, the South African bank ABSA, which had been engaged in lengthy negotiations to purchase a 49% shareholding in ZNCB, announced that it had withdrawn its bid. It is impossible to conclude conclusively that this directly resulted from the protests. However, popular opposition to the sale, and its likely consequences, appeared to have forced the Zambian Government to pressure potential purchasers to meet various social obligations, primarily the maintenance of unprofitable rural branches. This, and the lengthy negotiations that resulted, appear to have influenced the withdrawal of ABSA's bid.[31] The sale of ZNCB was effectively removed as a pre-condition for debt relief and it appeared unlikely that the other companies earmarked for privatisation will be sold in the near future. However, the ZPA issued a new call for expressions of interest in the purchase of a 49% share of ZNCB.[32] Mwanawasa has reiterated his opposition to any privatisation of ABSA that does not have 'a human face', apparently endorsing a Parliamentary vote preventing the sale of the 49% share to foreign interests.[33]

Whilst not a substantive break from neo-liberal policies, Mwanawasa's populist attempt to shift the blame for their consequences from his administration to external agencies reflected the rising level of popular discontent with the enduring dominance of Zambia's economy by foreign capital. The Government increasingly found itself squeezed between IMF strictures on spending and pressure from public sector unions demanding further salary increases. A strike by civil servants in mid-2003 was prompted by the Government's reneging on its collective agreement, a decision reversed after the IFIs pointed out that they were not included in the annual budget they had approved. As Finance Minister Ng'andu Magande explained, 'We are running the country but the budget is controlled by donors'.[34] The strike was declared illegal and union representatives were harassed by the police. The budget announced in January 2004 imposed a salary freeze on public sector workers, as part of efforts to remain within PRSP-imposed spending limits. This prompted a major demonstration and a public sector general strike in February 2004, to my knowledge the first ever called by the trade union movement.[35] The awarding of substantial debt relief in 2005 prompted calls by unions that the stringent controls on public sector wages be removed.

Mine privatisation, 1997 - 2000

The enduring centrality of ZCCM to Zambia's political economy was again demonstrated in the course of its privatisation. This process, completed in March 2000, was shaped at every stage by Zambia's dependent relationship with the IFIs and international mining capital. The process breached the requirements of Zambia's Privatisation Act for transparency in bidding processes, stakeholder consultation and regard for social and environmental impact. This influenced the failure of privatisation to realise significant revenue.[36] During this period, the IFIs sought to privatise the mines as rapidly as possible, regardless of the consequences for income generation, living standards and sustainability. MUZ publicly complained in 1998 that ZCCM and the Government were failing to meet legal requirements for consultation over the privatisation process.[37]

The Anglo American Corporation (AAC), with its 27.3% share in ZCCM, retained an explicit veto on the sale of the corporation's assets and was the obvious buyer should it be sold as a single unit. The Government was, however, reluctant to sell the mines to a single international corporation. Arguments over such issues initially delayed the privatisation process.[38] A World Bank-funded report recommended the unbundling of ZCCM into smaller companies and their sale through a competitive bidding process, with Government retaining a minority share in a rump ZCCM. The Government, the donors and AAC approved this process and ten packages were advertised for sale in February 1997.

Months later, Chiluba unexpectedly appointed a Privatisation Negotiation Team (PNT) headed by Francis Kaunda. Chairman and Chief Executive of ZCCM in the 1980s, Kaunda retained strong personal links with senior officials in international mining concerns, particularly AAC.[39] A bid for the largest package by four international mining companies (the 'Kafue Consortium') was rejected by the PNT in 1998. Donor anger at the resultant delay in privatisation led to the withholding of $238m of balance of payments support, at a time when ZCCM was making increasing losses. AAC subsequently benefited from the Government's resultant parlous economic position, acquiring Nchanga and Nkana mines for $18m.[40] As a major report on Zambian privatisation argues:

> … it was not until the company, from this position of strength, had negotiated and signed a memorandum of understanding with the Government over the purchase of Nkana, Nchanga and Konkola … that the donor community indicated that it would consider the approval and release of funds.[41]

AAC secured additional concessions to Zambia's already liberalised tax regime and shifted the costs of compensating 3,000 workers retrenched on the eve of sale to the Government. Notwithstanding these generous concessions, AAC divested itself of all its Zambian assets in January 2002, raising additional questions regarding the advisability of the sale.

Luanshya's Roan Antelope Mining Company (RAMCoZ) was the first major ZCCM asset to be sold. Francis Kaunda's PNT awarded RAMCoZ to the Binani Group of companies in 1997, disregarding legally binding tendering and consultation requirements. Kaunda's friendship with the Binani family appears to have influenced the decision.[42] Binani's promised investment of $69 million never transpired. The mine was asset-stripped and Binani failed to cover many of the costs it took on from ZCCM. $35 million of Binani funds, ostensibly to purchase the mine and to pay terminal benefits, disappeared. A former senior RAMCoZ executive confirms that the MMD received regular payments from the new owners, in *de facto* exchange for evading social responsibilities.[43] Binani also failed to meet its obligations to pay terminal benefits to retiring and retrenched workers.

When the local MUZ branch raised these issues with RAMCoZ in 1998, Branch Chairman Cameron Pwele was dismissed.[44] This prompted an eight-day strike, violently repressed by security forces, in which two people died. The strike achieved Pwele's reinstatement, but benefits remained unpaid. In October 2000, receivers were sent in to recover Binani's debts.[45] Mining operations were suspended in March 2001. For two years, retrenched miners, and those still employed to carry out essential services, went largely unpaid. In 2003, J&W Investments, a company based in Switzerland, signed a contract with the Government to take over RAMCoZ. Production in one part of the mine restarted in 2004, but only one-third of the mine workforce has been re-employed.

In a town that owes its existence to the mine, the impact of its closure was evident in every aspect of life in Luanshya during my research visits there in 2001-03. Most shops closed down. Transport services were virtually non-existent. The municipal council, which derived most of its income from the mine, was unable to provide basic services. Many miners' houses lacked electricity, water, and sanitation. Miners' families were unable to eat more than once a day and their children were excluded from school because of their parents' inability to pay fees.[46]

Cameron Pwele, retrenched by RAMCoZ in 1999, was elected as UNIP MP for Roan in 2001. He consistently clashed with the national MUZ's refusal to represent its members in regard to RAMCoZ and accused the union leadership of complicity in the corrupt privatisation.[47] He links this to MUZ's relationship with the MMD:

> They were supporting MMD. And they left the workers in the dark. They stopped thinking about the workers, because they thought they were Government and they could do anything. Anyone who could speak on behalf of the workers became an enemy.[48]

Many mineworkers are suspicious that MUZ officials benefited from the sale of ZCCM's housing stock. Chiluba won significant electoral support in 1996 by declaring that all ZCCM's 40,000 houses would be sold to tenants, mostly miners. However, senior ZCCM executives and some national MUZ officials acquired valuable houses at low prices.

Most miners primarily blame the Government and the new investors, rather than their union, for the problems of privatisation.[49] Some agree with MUZ leaders that the only alternative to privatisation was the closure of most mines.[50] Mineworkers generally express discontent with the specifics of privatisation, rather than generalised opposition to the policy.[51] For example, Boniface Kumwenda criticises the post-privatisation casualisation of mine labour:

> … most of the people now are being retrenched. And they're employing more contractors to do the job now. I think they are there for cheap labour. They don't want to own a lot of workers … there are people … getting K20,000 [$4 per day], K10,000 [$2 per day] … as contractors.[52]

A minority of former and present MUZ officials are highly critical of the union's role in the privatisation process:

> … the miner was not well informed about what was going on to the … privatisation … all they were getting [was] that there are new people who are coming in to take over the companies … it was engineered by the Government. But MUZ it didn't do enough homework. Try and put a checkpoint somewhere, as watchdogs.[53]

Frustration at the MUZ's inability, or unwillingness, to prevent a further real-terms decline in wages, have led to wholesale resignations from MUZ. Membership, which had fallen to 32,600 in 1997, is now less than 25,000 (no exact figure is currently available).[54] Across the Copperbelt, MUZ appears to be in terminal decline, losing significant membership to the rival National Union of Miners and Allied Workers (NUMAW), which was registered in 2004. Some miners argue, as their predecessors did in the UNIP era, that MUZ has been weakened because it is too close to the Government.[55] Percy Chanda:

> we were moving from a communism set up to a capitalist set up … those who were … in the government … they had their own interest. Which is far away from benefiting a worker … MUZ went in blind without understanding certain things. And that's why now the miner is almost in the cold … MUZ was a sub-Branch now of State House … Chiluba was hoodwinking them, and they were very willing.[56]

A return to militancy?

In 2005, subsequent to the completion of the primary research for this study, Zambia's copper mines and mineworkers again found themselves at the centre of national economic and political life. A dramatic rise in the international copper price made Zambia's mines profitable for the first time in 30 years. New mines were opening and significant exploratory development was underway in North-Western Province, optimistically described by President Mwanawasa as 'the new Copperbelt' in August 2005.[57] However, renewed profitability highlighted the historically low proportion of mine profits that, following mine privatisation, now accrues to Government revenue. It also prompted critical examination of the poor record of the new mine owners in labour relations, investment and safety.

In April 2005, 49 workers were killed in an explosion at the Beijing General Research Institute of Mining & Metallurgy (BGRIMM) explosives plant in the Copperbelt town of Chambishi. The plant, previously part of ZCCM, was sold to the Chinese company as part of the privatisation process. Days after the country's worst industrial disaster for 35 years, BGRIMM was unable to confirm the exact death toll, because most of those killed were casual workers whose names were not recorded. There was widespread anger at the fact that casual workers, paid between US$15 and US$30 per month, could be employed in such a hazardous working environment. Government Ministers criticised the company and the Ministry of Mines froze the issue of any new mining licences. A national day of mourning was declared to mark the mass funeral of the deceased. At the funeral, family members of those killed expressed their anger at both BGRIMM executives and Government Ministers. Police beat back protestors attacking the Government's failure to properly regulate privatised companies.[58] It is widely suspected that new owners are able to evade safety controls by bribing government inspectors.

Popular anger was also expressed at the Government's sale, in late 2004, of the majority share of Konkola Copper Mines (KCM – the former AAC holdings) to Vedanta Resources for just US$25m, an amount it recouped in its first three months of operations.[59] In July 2005, KCM workers demanded a 100% pay rise and took unofficial strike action in support of their claim, explicitly linked to the company's substantial profits. When both MUZ and NUMAW leaders agreed to KCM's offer of a 30% pay rise, mineworkers rioted in protest. Michael Sata, leader of the opposition Patriotic Front (PF), in offering the striking miners his support, demonstrated in his actions (and in the Government's response) the continuing political importance of Zambia's mineworkers. Sata, whose increasing popularity is based on his opportunist attacks on IMF-influenced economic policies, was subsequently detained and charged with sedition for inciting miners to riot.[60] The dispute also led to the removal from office of MUZ's President and General Secretary, for having incorrectly claimed that KCM's offer of 30% was final.[61]

In the run-up to Parliamentary and Presidential elections due in 2006, opposition parties can be expected to seek political capital from the widespread unpopularity of foreign mining investors. Whether Zambia's remaining mineworkers can harness an apparently renewed militancy to recover some of the losses they have experienced may depend on the lessons they draw from their recent collective experiences of economic and political change.

Mineworkers and political change

In 2002-03, I asked former and present MUZ officials to reflect on MMD governance and whether in hindsight they would repeat their support for the movement for the pro-democracy movement in 1990-91. Whilst the majority of respondents saw the impact of most MMD policies as negative, the sale of houses to miners has some support:

> One of the most good things MMD did is to sell property … Most of the people, what they used to do when they had retired, especially miners, was to go home [to their villages of origin]. When they went home they never even stayed two, or one month; some of them died … They died not because they were witched, no. They died because of the environment they found was there. Because some of the people were born here … and they are used to this environment … typical villages where you … have to walk distances to grow [crops] … They'll be walking distances to draw water. To come and bath … but they're used to be bathing every day. It's difficult. So, that's one thing the MMD has done … for the people.[62]

Many respondents reported improvements in public transport and food supplies.[63] Evelyn Musonda:

> MMD might have its own problems but compared to what we went through during the UNIP era, I think they have done much better. Transport wise you can see … there are a lot of vehicles now … there were times [under UNIP] that I used to spend five days at the [bus] station just getting to Kabwe…[64]

Boniface Kumwenda disagrees:

> I never saw anything changing. The things [are] there in the stores, in the shelves, but then for a person to go and buy, he had no money … he wants to eat but he cannot go and get those things … it's no different to not having things on the shelves…[65]

David Longwani expresses the views of a significant minority of mineworkers:

> we have realised that the support was a wrong one. In the sense that, there're no salary increments from the time the MMD people came in … Actually we're now regretting, they should have just supported UNIP. Because what has happened now that after privatisation of the mines … the *wazungus* [whites] that have come in now, they are not even looking at people's plight.[66]

George Mulenga specifically relates the labour movement's support of the MMD to subsequent problems.[67] John Bwalya agrees that, 'the Government has swallowed all the powers of the MUZ … it's just left toothless.'[68] Patrick Kanyanta, who now campaigns for the payment of terminal benefits to retrenched mineworkers, nevertheless believes MUZ and ZCTU were right to support the multi-party movement:

> Now, what was wrong was now, when after giving this support, the Government would not go back to consult the members or the workers on what next … the time Frederick Chiluba became President, the top leadership from the Mineworkers' Union, ZCTU, they never pushed with that zeal … the checks and balance there went away, they became more partisan.[69]

Raphael Mwanza expresses the commonly held view that it was right to support the MMD in order to oust UNIP and to achieve multi-party democracy:

> MUZ was supposed to give that support for one reason. Just to change the UNIP regime. For that reason, the support was genuine. But mistakenly, people blindly thought the support was because, we're going to improve our lives. Which proved not to be forthcoming … As it is now, it's like from one-party state to another one-party state.[70]

Pascale Mwansa agrees:

> It was important though to support MMD, because [otherwise] we wouldn't have come out from the one-party state. That was a very

> good move. But the people entrusted with power are the people who subverted it.[71]

Oswell Munyenyembe, now MUZ Vice President, draws a regional comparison to argue that this subversion could have been lessened by a more contractual relationship with the MMD:

> there's no regret of us reaching [changing] the Government. I think what we lacked was checks and balances ... when we change government, would [should] have, just like our colleagues have done in South Africa. They are supporting the ANC, but they will say, we are going to have a strike. What you are doing is wrong. That's what we were lacking here.[72]

Some interviewees argue that the key achievement of the pro-democracy movement was free speech:

> In a way, we've had a raw deal, the labour movement. But generally, the change was worth it, because it has opened a lot of people's minds. People know their rights, and they can stand up and speak. They can criticise the politicians which was unheard of before the 1991 elections...[73]

I also asked mineworkers what they believed was the appropriate relationship between trade unionists and political parties in general. Most now believe it is important for the union to stay out of 'direct' involvement in politics, although they define this in varying ways. A K Nguluwe:

> ... as trade unionists, our role is to ... correct the Government, criticise where there is need to criticise ... By us going to be [in] politics, we are destroying the essence, where the labour movement was founded ... even us as Mineworkers' Union of Zambia, we should not even support the Government of the day, its dangerous, because we are the voice of the people...What we can just do, like what we did in 1991, [is to] support a change ... Immediately that change comes, we come out...[74]

Edward Kabundi's views are widely shared:

> … a unionist should keep away from politics. He should not mix with politicians. Because immediately he mixes with politicians, then immediately his union starts becoming weaker and weaker.[75]

Boniface Kumwenda, never an MMD supporter, feels vindicated by subsequent events:

> I was against that announcement by the National Chairman [in 1990], to say we are in support of MMD … when you say that, you tie yourself to their policies … they should have said we are going to remain watchdogs of any political party which comes into power. And see what it's doing to the people.[76]

Albert Chali places these arguments in a historical context:

> … it was not possible for UNIP … to defeat the colonial government had it not been for the support from the labour movement. Similarly, even after UNIP was under one-party state, Chiluba … did make some effort because of the support from the labour movement. But in most cases it's dangerous, if you put all your efforts together with the government of the day. Because once those people start making some political blunders, it will be very difficult for you to challenge them … But there's no way you can say, the union should separate from this, no. It's not possible.[77]

David Ndalama agrees that political participation is unavoidable.[78] Timothy Mfula distinguishes between political engagement and political manipulation:

> … you cannot separate unionism from politics. Because unionism itself is politics. But what we should do maybe is draw a line between the general politics and the labour movement politics. If we have to go into an alliance with a political party, the Union should be able to give what it expects from the political party if it comes into power … But for us, unfortunately, we went into bed with the politicians with a blank cheque. We never gave them conditions and they were able to ride on our backs. And to make matters worse, the people that were in power were unionists, so, we have no cause to complain, we reaped what we have sown.[79]

Conclusion

This chapter has explored the way in which economic liberalisation and political governance under the MMD failed to address the expectations that prompted Zambia's mineworkers to participate in the campaign for multi-party democracy in 1991. It has demonstrated that, despite formal democratisation, unionised workers continue to lack effective political influence. This is partly because many unionists believed that a government led by a former union leader would effectively represent their interests. It is also because workers' negative experience of state ownership of industry under UNIP led many of them to accept the argument of politicians that the removal of state control would improve their working conditions. Zambia achieved political liberalisation at a time when its economic weakness made the new government particularly vulnerable to international pressures to radically liberalise its economy. Organised workers were unprepared for the unpropitious economic and political environment that led the MMD to implement policies that were injurious to their living standards and employment prospects, and the labour movement lacked the capacity to generate a clear oppositional response. Notwithstanding their marginalisation from formal consultation processes, the enduring loyalty of union leaders to the party they had helped create and the vested interest of those leaders in state-based patronage, prevented the effective expression of working-class discontent. After more than a decade of MMD rule, the ZCTU and the wider labour movement began to rediscover the autonomy from political interests that was the basis of its effectiveness under UNIP. It again mobilised against the impact of economic liberalisation policies and the politicians implementing them.

For Zambia's mineworkers and their union, this appears to have come too late. As has been outlined, the process of mine privatisation was designed to meet the demands of Zambia's creditors rather than the needs of her people. This resulted in the (temporary or permanent) closure of some mines, a consequent significant reduction in foreign exchange earnings from copper mining and the retrenchment of more than 50% of Zambia's mineworkers. This resulted in the immiseration of mineworkers' families and communities and the virtual collapse of the Copperbelt as a viable urban region.

Mineworkers remain bitter that their personal actions and sacrifices in the movement for multi-party democracy have achieved so little. As this chapter has shown, whilst some believe in retrospect that it was wrong to support the MMD *as a movement*, a narrow majority of my respondents believe this was necessary to ensure political change in Zambia. They are however divided as to whether it was necessary to support the MMD *as a political party*. Mineworkers express greater unity in arguing that, in general, trade unions and their leaders should not become directly involved in politics, equating this with the inevitable weakening of trade union organisation. Most believe, however, that the labour movement has an important 'watchdog' role over political parties. Many acknowledge the political nature of industrial relations, particularly in an industry that is once again central to the relationship between the state, political power and economic resources in Zambia. The dilemma of the labour movement's role in influencing political change is one with which mineworkers and the wider working-class, will continue to struggle in the future.

CONCLUSION

This study has examined the actions and consciousness of Zambia's mineworkers. It has sought to explain why they and their union have played a diverse range of important and influential political roles in the changing political and economic context of post-colonial Zambia. It has rejected established explanations, dominated by nationalist and developmentalist thinking, which have generally failed to explain or sometimes even to recognise that the union and its members have had a substantial political influence since Independence.

Zambia's place in the global economy

This book has sought to demonstrate the enduring significance of the way in which Zambia was, and remains, integrated into the global economy solely as a provider of a single raw material, copper. Northern Rhodesia's urbanisation, exceptional in sub-Saharan Africa, was both necessitated and enabled by mine revenues. Multinational mine companies, the colonial state, white settlers and the African mine workforce competed over the distribution of copper revenue. However, its benefits accrued to African mineworkers only when they actively contested the existing level of wages and services provided by the mine companies. Mineworkers, understanding the centrality of their labour to the profitable exploitation of mineral wealth, believed this entitled them to a larger share of that wealth than mine companies were otherwise prepared to grant and periodically withdrew that labour in order to influence its distribution in their favour.

At Independence, Government economic policy assumed that copper mine revenue would fund post-colonial development. Government ministers and development advisors therefore believed that copper revenue conceded to Zambia's mineworkers was lost to this project. This belief informed the early clashes between the post-colonial state and mineworkers, as the latter sought to realise their aspirations of Independence by acquiring a larger share of the mine revenue that their labour enabled. Zambia's failure to attract foreign investment only increased its dependence on copper revenue and encouraged increased state control of the economy. As Hamalengwa was the first to identify,

President Kaunda's announcement of the 51% nationalisation of the copper mining industry had, as one of its aims, the increased control over mineworkers' labour by the state.[1] For Burawoy's respondents, tripartite cooperation between the state, the nationalised companies and their union was designed to ensure their more effective exploitation.[2]

Nationalisation failed to deliver increased investment in, or revenue from, the mining industry. When the copper price fell in the mid-1970s, the developmental assumptions of the post-colonial state collapsed with it. Organised labour could extract wage concessions in the 1960s through industrial action; this became increasingly difficult to achieve. Developmental advice now came from the IFIs, which proposed systematic reductions in the provision of social services that, for many Zambians, represented the primary achievement of Independence. Neo-liberal economic advisors sought to reduce state and parastatal expenditure in general and salary budgets in particular. Notwithstanding a rhetorical commitment to economic diversification, copper revenue was a preoccupation of the IFIs. Despite its lack of profitability, ZCCM remained central to economic policy throughout the 1980s. UNIP utilised ZCCM's foreign exchange earnings to fund projects designed to enhance its political prestige and the personal enrichment of its leaders. The accrual of such earnings to the state prevented the reinvestment necessary to ensure the copper mines' international competitiveness, undermining their long-term viability.

The subsequent liberalisation of the Zambian economy only demonstrated the failure of post-colonial governments to overcome its global marginalisation. UNIP established state-owned industries in the late 1960s precisely because private investment was not forthcoming. Privatisation did not lead to significant investment, leading instead to the closure of much local manufacturing and industry and the replacement of its output with imports that are unaffordable for the vast majority of Zambians. The underlying decline in the mining industry's competitiveness has been revealed by its privatisation. In addition, the recent rise in copper prices has not led to significantly increased income to the state, because of the historically low tax rates conceded to the newly privatised mine companies. For the first time in seventy years, the Zambian state is no longer dependent on copper mine revenue, but nothing (except dependency on foreign aid) has replaced it.

The state and the union movement

A second major argument of the study has been that the relationship between the UNIP-dominated state and the labour movement was primarily informed by the desire of the former to effectively control the latter. This was done, as in the colonial era, by the manipulation of union organisation and leadership, as a tactic to control union members. UNIP, like the colonial government, searched in vain for a responsible labour movement that would persuade workers to delay the improvements in wages and conditions that they expected Independence to deliver.[3]

The early passing of the Trade Unions and Trade Disputes Act and the legislative establishment of the ZCTU demonstrated the importance attached to this issue. The subsequent Industrial Relations Act (IRA) of 1971 was publicly portrayed as enabling workers' participation in the economy. In practice, President Kaunda sought to establish a powerful union organisation, the primary aim of which was the control of workers by responsible union leaders who would be loyal to UNIP. Corporatist aspects of the IRA, such as Works Councils, did not significantly dilute management authority. The appointed leaders of the ZCTU, particularly President Wilson Chakulya, demonstrated loyalty to UNIP, to the disquiet of many national and local union leaders. In 1971, at their first opportunity, they elected as Chakulya's replacement a union leader critical of UNIP. Newstead Zimba, and subsequently Frederick Chiluba, led the ZCTU in a new direction. Utilising the powerful centralised Congress organisation enabled by the IRA, they increasingly challenged UNIP's economic policy, particularly as it affected organised workers.

Chiluba understood that the mineworkers occupied a strategic position in the national economy and that their discontents could be utilised to challenge the policies, and subsequently the structures, of the one-party state. During the strikes of 1981, mineworkers demonstrated this potential in providing an effective challenge to the detention of ZCTU leaders. In the 1980s, the ZCTU leadership's direct links to MUZ branch activists provided an alternative organisational base which, by the late 1980s, was capable of mobilising politically active mineworkers across the Copperbelt. Chiluba's position in the trade union movement enabled him to effectively express the various discontents of Zambians with the impact of structural adjustment and other UNIP policies. By the time he called for the return of multi-party democracy in December 1989, most mineworkers already identified Chiluba as their most effective representative.

During the period when Chiluba was simultaneously Chairman General of the ZCTU and a leader of the multi-party movement that became the

MMD, the resources of the ZCTU were placed at the disposal of the movement. However, the labour movement's importance lay not only in its organisational and financial contribution, but also in its potential capacity to mobilise mass protest and industrial action. Kaunda's failure to repress the MMD stemmed from awareness that the party's supporters in the labour movement had the capacity to make Zambia ungovernable.

The mineworkers' union

The MUZ (and its predecessors) cannot be understood as a homogenous organisation that acted in a unified way, reflecting the aspirations of its members. The union was at the centre of the continuous conflict over the control of mining and labour resources. Its leadership came under contradictory pressures. On the one hand, UNIP, like the colonial state, equated responsible union leadership with a willingness to control and discipline union members. On the other hand, rank-and-file members refused to relinquish control of a union that they defined and utilised as the collective representative of their aspirations. The MUZ was simultaneously the Katilungu House officials with their direct links to State House and ZCCM senior management; the branch officials who contested their dominance at Supreme Council and Biennial Conferences; the shop stewards challenging workplace controls on behalf of members; and the members themselves, commonly dissatisfied with the limited achievements of all these leaders and (with their wives) prepared to challenge them in the contested space of the public meeting. It may indeed be argued, in distinction to the traditional slogan of the international labour movement, that the mineworkers' union's evident and enduring disunity has been its greatest strength.

At Independence, leaders of the ZMU, whilst generally supportive of UNIP, nevertheless sought to maintain the union's autonomy from the party and its new government. This was evidently unacceptable to UNIP: via the United Mineworkers' Union in 1963-64, and the attempted manipulation of ZMU elections in 1965, the ruling party unsuccessfully sought to secure a union leadership that would be unambiguously loyal. The strikes of 1966 intensified the urgency of establishing effective control over Zambia's mineworkers. The Ministry of Labour influenced the merger of the ZMU with other mining unions, to form the MUZ. The new MUZ leadership was significantly more supportive of Government policy than their predecessors, consistently opposing attempts to organise industrial action and establishing an apparently rational and technical approach to industrial relations, exemplified by the 1970 Job Evaluation Agreement.

It was, however, this Agreement's failure to meet mineworkers' enduring expectations of a post-Independence transformation in their working and living conditions that sparked challenges to the leadership of MUZ President David Mwila. These challenges, primarily based on industrial discontent, were nevertheless perceived as a threat by both the nationalised mine companies and the state, which intervened to prevent them in 1971 and again in 1974. UNIP's willingness to intervene directly in MUZ's internal affairs appeared to have delivered the mineworkers' union that it had always sought. By the late 1970s, however, the Mwila leadership came under increasing pressure, as it failed to maintain the real value of wages and to reflect mineworkers' grievances. Timothy Walamba, a leader initially willing to support industrial and community-based action to address such grievances, replaced Mwila in 1982.

Walamba's leadership was, however, increasingly squeezed between a one-party state whose attempts to manage a growing economic crisis led it to demand increasing sacrifices from Zambia's people, and members whose unwillingness to accept the inevitability of lower living standards led some to take (sometimes violent) action that necessitated direct conflict with state forces. The inherent conflict within MUZ was highlighted in the 1985 Mukuba pension strike. It was expressed more generally in clashes between the union leadership and branch leaders who more closely reflected the aspirations of the mineworkers amongst whom they lived and worked. UNIP's 'one union, one industry' approach, arising from its attempts to control the labour movement, provided a powerful deterrent to the creation of splinter unions. Instead, dissenting mineworkers concentrated on capturing the MUZ from its leadership. Such attempts were however initially thwarted by the lack of democracy within the union.

UNIP's enduring need to control Zambia's mineworkers was exemplified by the appointment of Walamba and MUZ General Secretary John Sichone to its Central Committee in 1988. It was believed that separating the MUZ leadership from the oppositionist ZCTU would weaken the latter's capacity to challenge the policies and prerogatives of the one-party state. Mineworkers ousted Walamba in 1990 because they rejected his affiliation to UNIP and the perceived negative impact of this on their living conditions. Jonathan Simakuni was elected to the MUZ Presidency because mineworkers believed that a union leadership opposed to UNIP would more effectively advance working and living standards.

However, with its declared support of the MMD, the new MUZ leadership ceased to come under pressure from its members, most of whom now believed that political change, rather than autonomous industrial militancy, would address their social and economic aspirations. In the early 1990s, the MUZ leadership was closely affiliated to the new Government and to President Chiluba in particular. Whilst subsequent disillusionment with the MMD and the negative impact of privatisation have increased discontent with the union's leadership, the devastating effect of mass retrenchment on the mining communities has militated against successful challenges to it.

Mineworkers' actions and consciousness

Understanding the various roles played by the mineworkers and their union in post-colonial political change requires an appreciation of the values held by mineworkers, in relation to their workplaces, communities, and place in Zambian society. When one asks mineworkers to justify their demands for improvements in wages and conditions, they will commonly respond with the challenge, 'Have you ever been down a mine?' Zambian mineworkers believed, like mineworkers elsewhere, that their work was uniquely harsh, dangerous and important and deserving of fair reward. A 'fair wage' was defined not in terms of the limitations imposed on the Zambian economy by its subordinate global position. Instead, it was measured in relation to the wages paid for similar work in other countries, to the profits made by mine companies and the way those profits were consumed by private and state capitalists. This definition implicitly offered a moral challenge to the national and international distribution of wealth and resources. Mineworkers were not, as much of the established literature has suggested, an apolitical labour aristocracy, concerned only with improving their own living conditions through an 'insider' relationship with the mine companies and the post-colonial state. Nor were they, as Bates in particular claimed, generally supportive of UNIP and the one-party state.

During the different periods examined by this study, the political actions of mineworkers have at times been more consciously political and influential than at others. Much of the time, most miners were preoccupied with the politics of daily economic survival. At all times times, however, their political actions and consciousness were informed by their values, and their powerful sense of collective identity. Indeed, the consistency of mineworkers' aspirations is striking. From 1940, when the demand for equal pay for equal work was first made, to Simakuni's demand for an end to racially-based wages in 1990, Zambia's mineworkers have over a half-century refused to accept the necessity of sacrifices demanded by colonialists and capital, nationalists and developmentalists, donors and democrats. This obstinacy has been variously understood as aristocratic, selfish, elitist, apolitical and politically conservative. By rejecting such labels in favour of an empirical exploration of mineworkers' actions and their motivations for them, this study has sought to demonstrate that mineworkers make sense of their realities through a set of enduring values that are in fact the antithesis of these characterisations. These values include: an aspiration for relative equality of consumption and of sacrifice; a demand for the adequate valuation and compensation of hard and hazardous work; an expectation of the public accountability of political and labour leadership to their constituencies; and a desire for natural and human resources to be utilised for the improvement of society as a whole. These values have been shaped by mineworkers' own experiences of post-colonial societal and political change. Whilst not expressed in explicit socialistic or communistic ideologies, the language of which was appropriated in the period under study by UNIP, they nevertheless amount to a form of working-class consciousness which bears comparison with that of mineworkers across the world.

These values were consistently reinforced by mineworkers' daily workplace experience, by the continuity of workplace and community in the mine townships and by a shared belief that collective direct action by the mine workforce and community was necessary to achieve improvements (or resist reductions) in their conditions. Similar values appeared to be held by the wives of mineworkers (although this study presents only limited evidence in this regard). Union activities took place as much in the township as in the mine. Public meetings enabled the effective expression of the aspirations and grievances of the mining community. They also provided opportunities to challenge and assess union leaders against implicit criteria of good leadership. These were, essentially, bravery in the face of management and state intimidation; a principled resistance to bribery and other material gains made possible by their office; and a willingness to speak honestly to the community regardless of the consequences. Jonathan Simakuni exemplified these values and continues to do so because his early death prevented any possibility that he might stray from his principled path, as other MUZ leaders are perceived to have done. Whilst this emphasis on the morality of individual leaders checked their co-option and corruption, it may have militated against a more politicised analysis of the social position held by union leaders, limiting in particular mineworkers' capacity to critically analyse Chiluba's transition from ZCTU Chairman to State President.

This study has sought to follow Edward Thompson's principle that the consciousness of workers can only be understood in the dynamic context of historical change.[4] Mineworkers, like other Zambians, had particular (and variegated) aspirations for national Independence. They expected an end to the racial inequality and violent supervision that lay at the heart of colonial industrial relations. Most also expected an extension of the practical authority of union representation that they perceived was necessary to ensure significant improvements in wages and conditions, funded by the redistribution of mine company profits that their labour generated. Whilst UNIP shared the mineworkers' aim of advancement into jobs previously reserved for whites, it did not accept that this would mean equality of pay. In an initial period of post-Independence economic growth, mineworkers were in 1966 able to secure significant wage increases, in the face of opposition from mine companies, and despite the detention of branch leaders. Mineworkers found that the post-colonial state was as opposed to independent industrial action as its colonial predecessor and was willing to use similar methods to prevent it.

Mineworkers initially voted for UNIP because they believed its promises of radical post-colonial change. When these proved for some mineworkers to be inadequate, many supported the UPP. Whilst not providing a clear ideological alternative to UNIP, the UPP expressed many of the discontents of mineworkers with the post-colonial economic settlement. Understanding the strategic position of mineworkers in the Zambian economy, the UPP prioritised the mobilisation of their support. Whilst the declaration of the one-party state addressed this challenge, the repression of dissent delivered only temporary quiescence. Rank-and-file organisation, and the willingness of mineworkers and their communities to challenge the implementation of UNIP policy, ensured that the mine townships remained a locus of resistance to the UNIP state's attempts to achieve effective control over Zambia's economy and society, and over workers through the agency of the union bureaucracy.

The reduction in strikes in the 1970s was achieved primarily because of increased repression, not because mineworkers were convinced that their grievances were being addressed. Whilst the state was able to limit industrial action at the point of production, the expression of discontent was partially displaced to the mine townships, which remained outside effective state control. Grievances were expressed by miners, but also by their wives, whose lack of representation paradoxically freed them to play a leading role in representing their communities' aspirations. UNIP's attempt to establish effective control of the mine townships, via the 1980 Local Administration Bill, prompted Zambia's most significant post-colonial strike in January 1981. Mineworkers sought to prevent the replacement of elected Mine Township Councils with District Councils dominated by Presidential appointees and the decline in their living standards they expected would result from this. Whilst this decline did indeed take place, the concessions granted by the state represented a victory. Subsequent industrial action by miners and their wives achieved its aim of subsidised mealie meal sales by the mine companies. Later that year, the threat of industrial action contributed to the early release of detained ZCTU and MUZ leaders.

In the 1980s, some of Zambia's mineworkers and their families used industrial and community-based direct action to challenge their declining living standards and to resist attempts by UNIP to undermine and/or gain control of their union. In so doing, they expressed their discontent with the consumption by a corrupt state-based elite of the surplus value produced by their labour. The 1985 Mukuba pension strike achieved its immediate aim. The 1986 food riots, which began in the mine townships, were decisive in undermining the IMF-designed structural adjustment programme. Mineworkers were however unable to prevent a significant decline in their living standards. The limitations of industrial action prompted a significant number of mineworkers to engage in underground political opposition, partly co-ordinated by the ZCTU leadership.

In 1990, the grievances of most mineworkers found expression in the MMD. Political democratisation appeared to offer what national Independence had failed to deliver: an end to racial inequalities in wages; a fairer distribution of the income generated by the mines; and the democratic space within which they and their union could effectively express their aspirations. Their hopes in this regard were based in large part on the labour movement's central role in the MMD, symbolised by Frederick Chiluba's leadership. This was not, however, reflected in the effective representation of mineworkers' interests in the party.

Mineworkers' practical experience led most to believe (wrongly) that the removal of state intervention would improve both ZCCM's fortunes and their own wages and conditions. The failure of these hopes to be realised led most mineworkers to believe, as did those who resisted UNIP control before and after Independence, that the direct involvement of trade unions and their leaders in politics tends to weaken union organisation. This preference for activity within the union and workplace as a method for advancement, rather than affiliation to a political movement, echoes the syndicalism of other working communities in other times.

In all of this, the impact of structural factors in defining Zambia's subordinate position in the global economy, the resultant weakness of indigenous private capital and the consequent model of state-led economic development adopted from the late 1960s, is undeniable. In the long run, Zambia's mineworkers struggled against the declining value of their industry to the world economy and ultimately did so in vain. The structural problems preventing mineworkers realising their aspirations increased greatly over time. In the early 1990s, the political democratisation that they hoped would enable their aspirations to be addressed was immediately undermined by the parlous economic position of the new multi-party state.

Mineworkers' struggles are nevertheless instructive in what they tell us about way in which external constraints were experienced by a key section of Zambian society. Mineworkers refused to accept the declared 'realities' of development that resulted from the unwillingness of the post-colonial administration to challenge its lack of control over its primary economic resource; its consequent (albeit uneasy) collaboration with international mine capital; and the limits this placed on social and economic change. Their actions, whilst not always driven by explicitly political aims, provided a consistent challenge to the political decisions that underlay the failure of post-colonial development and the rising poverty that has been the unhappy reality for most Zambians for much of the last forty years. By maintaining the autonomy of local union and community organisation, mineworkers were able to significantly contribute to the return of multi-party democracy. While this achievement did little to improve the lives of mineworkers, and may have paved the way for the (apparently irreversible) decline of their industry, the democratic space won by the multi-party movement cleared the way for civil society's current challenge to the implementation of privatisation demanded by the IFIs in exchange for debt relief. If the labour movement is to play a significant role in Zambia's future, it will be built upon the legacy of mineworkers' considerable contribution to post-colonial political change.

Labour and political change in post-colonial Africa

In an African context, Zambia's Copperbelt is in many respects exceptional. The high level of early and (until recently) sustained urbanisation; the national dependency on mineral production requiring the concentration of large numbers of relatively skilled workers; the relative absence of ethnic (though not racial) tensions within the mine workforce and communities; and the country's virtually unparalleled economic decline. These factors emphasise the unique nature of this case study. Nevertheless, there are lessons that can be drawn which might usefully inform our understanding of the relationship between organised labour and political change in post-colonial sub-Saharan Africa.

Firstly, attempts to extrapolate the consciousness of workers from their union leaders must be rejected in favour of a close examination of the dynamics within trade union organisation and the problematic relationship between leaders and members. This relationship is commonly conflictual, arising as it does from the inherently ambiguous position of union leaderships, whose role is both to express their members' demands and to limit those demands to those acceptable within the existing economic framework. The leadership of Zambia's mineworkers' union achieved an advance in their personal material conditions that distanced them from the day-to-day concerns of their members. In their control of union resources, and their use of patronage and bureaucratic mechanisms to maintain their positions of authority, MUZ leaders behaved in ways that largely reinforce Michels' 'iron law of oligarchy' and wider theories of the labour bureaucracy.[5] Mineworkers successfully challenged and removed such leaders, whilst not changing the inherent power relations in the union.

Equally problematic are commonly held assumptions about the nature of post-colonial states. Social scientists have frequently failed to engage with the challenging realities of the form of political independence achieved in sub-Saharan Africa. Nationalist politicians mobilised widespread support because they promised that Independence would enable the realisation of the aspirations of sections of African societies for their (variously defined) advancement. Such aspirations, when they lay outside the development plans drawn up by nationalists and their foreign advisors, were subsequently dismissed as unrealistic and illegitimate. In practice, official plans have been shown to be generally unrealistic in both their aspirations and methods for achieving post-colonial development.

Whilst the decline in the international copper price had a particularly dramatic impact in Zambia, the adverse turn in the terms of trade in the mid-1970s had similar effects in many African countries. Disillusionment led to a decline of popular support for nationalist parties; some of these were overthrown by coups with initial popular support, whilst others responded to the crisis in their legitimacy and the electoral threat of opposition parties by the establishment of one-party states. The debt burden led in many cases to the implementation of structural adjustment programmes by undemocratic governments. The negative impact of these unpopular policies generated significant popular support for pro-democracy movements in the early 1990s. The renewed demand for democracy, like the struggle for political Independence thirty years earlier, was for most Africans a means to an end, the fuller realisation of their aspirations for the advancement of themselves and their communities. Thus, in order to understand the history of post-colonial Africa, it is necessary to re-legitimise the diverse aspirations of Africans themselves.

This study supports recent international labour historiography in suggesting that the basis of workers' struggles lies as much in the community as the workplace. Burawoy's seminal study provided evidence that changes in the production process had a significant influence on industrial relations. My research, whilst not denying the importance of struggles over workplace control, has found that the major conflicts in Zambia's copper mines since Independence took place over wages and conditions, in particular the 'social wage'. In interviews, mineworkers primarily reported grievances regarding wages and living conditions, rather than their alienation at the place of production. Research has highlighted the central role of women in the wider struggles of the mine communities. Post-colonial labour studies need to ensure that community-based demands and aspirations are examined as closely as those put forward in the workplace. The common experience of mineworkers internationally, of overlapping community and workplaces resulting from companies' construction of dense housing projects adjacent to mines, has made understanding the complex interaction between these experiences central to the contemporary practice of mine labour history. Campbell's seminal study of the Scottish miners sought to address this problem by presenting two separate volumes focusing respectively on community-related issues and on trade unions and politics.[6] This study finds no such clear distinction between the two, demonstrating instead the way that particular events were played out in the interaction between ministry offices, air-conditioned company boardrooms, cramped underground excavations, and the open spaces of the township meeting place.

The study also demonstrates the limitations of an approach to labour studies that is based on a reified and idealised working-class, shorn of all alternative identities, which is cast by the author in an ideologically ascribed historical role. Rather, it is necessary to study the complex and multi-faceted nature of the working-class that actually exists. As Ferguson identified, Zambia's mineworkers never lost their connections with rural kin.[7] For many decades, they were significant providers of investment into their areas of origin, achieving influence and status in villages to which most retired.[8] Whilst few miners are now able to send funds to villages, and many are now able to retire in towns, the political influence of Copperbelt workers, based on their interaction with their areas of origin, continues to shape national politics.[9]

Mineworkers, because of their strategic position in the Zambian economy, were better able to express their discontent with post-colonial political change than most other sections of society. Their demands for improved wages and conditions necessarily had a political impact.[10] Mineworkers' powerful sense of their political influence was borne of experience, which informed and is informed by the enduring values set out above. This study has sought to understand the nature of that consciousness in relation to political change, rather than measure it against an external ideological benchmark. Approached in this light, researchers can make sense of the enduring influence of organised labour in pro-democracy and anti-liberalisation campaigns in Africa.

Finally, no historian can disregard the present social context in which s/he works. The workplace and social organisation of Zambia's mineworkers and the particular form of class consciousness that arises from it, is under unprecedented threat. The halving of mine employment, the drastic decline in the real value of mine wages (in particular the social wage that privatised mine companies are unwilling to provide) and the consequent dependency of mineworkers' families on alternative sources of income, have all weakened mineworkers' sense of solidarity. The sale of ZCCM houses, whilst providing retired miners with the permanent urban base so many desire, has resulted in an influx of non-miners into the townships and a consequent dilution in the shared experience of work and community on which the peculiar consciousness of mineworkers was constructed. The uncertain future of the mining industry has undermined the assumptions that underlie much labour and social historiography, i.e. that urbanisation and social and economic progress are inevitable processes that will continue to unfold. Rather, there is the distinct possibility that this history of the significant role played by Zambia's mineworkers and their union in post-colonial political change will prove to be an epitaph for a way of work and life that may soon disappear.

NOTES

Introduction

1 *Sunday Post*, 8/2/2004.

2 See, for example, H.S. Meebelo, *African Proletarians and Colonial Capitalism*, Kenneth Kaunda Foundation (Lusaka, 1986); J. Pettman, *Zambia: Security and Conflict* St. Martin's Press (New York, 1974).

3 R.H. Bates, & P. Collier, 'The Politics and Economics of Policy Reform in Zambia', in R.H. Bates & A.O. Kreuger (eds.), *Political and Economic Reform: Evidence from Eight Countries*, Blackwell, Oxford and Cambridge MA (1993), p.429.

4 See, for example, W. Tordoff (ed.) *Politics in Zambia*, Manchester University Press (Manchester, 1984).

5 A. Campbell, *The Scottish Miners, 1874 – 1939. Volume One: Industry, Work and Community*, Ashgate (Aldershot, 2000), p. 4. Campbell's definition is based on his reading of M. Savage and A. Miles, *The Remaking of the British Working Class, 1840 – 1940*, Routledge (London and New York, 1994).

6 M. Burawoy, *The Colour of Class on the Copper Mines*, University of Zambia Institute of African Studies (Lusaka, 1972); J. Ferguson, *Expectations of Modernity: Myths and Meanings of Modern Life on the Zambian Copperbelt*, University of California Press (Berkeley, 1999).

7 Key works include A.L. Epstein, *Politics in an Urban African Community*, Institute for African Studies University of Zambia/Manchester University Press (1958, 1973); J.C. Mitchell, *Cities, Society and Social Perception: a Central African Perspective,* Clarendon Press (Oxford, 1987); G. Wilson and M. Wilson, *The Analysis of Social Change Based on Observations in Central Africa*, Cambridge University Press (Cambridge, 1945).

8 Ferguson, *Expectations of Modernity*, passim.

9 V. Jamal & J. Weeks, *Africa Misunderstood: or Whatever Happened to the Rural-Urban Gap?*, International Labour Organization, Macmillan (Basingstoke, London, 1993), pp. 75-87.

10 B. Freund, *The African Worker*, Cambridge University Press (Cambridge, 1988), p. 18.

11 Ibid., p. 20.

12 F. Cooper, Decolonization and African Society: the Labour Question in French and British Africa, Cambridge University Press (Cambridge, 1996).

13 F. Cooper, 'The Dialectics of Decolonization: Nationalism and Labour Movements in Postwar French Africa' in F. Cooper and A.L. Stoler (eds.), *Tensions of Empire: Colonial Cultures in a Bourgeois World*, University of California Press (Berkeley, 1997).

14 E. Berg, & J. Butler, 'Trade Unions', in J. Coleman and C.G. Rosberg (eds.) *Political Parties and National Integration in Tropical Africa*, University of California Press (Berkeley, 1964).

15 D.C. Mulford, *The Politics of Independence, 1957–1964*, Oxford University Press (Oxford, 1967), p. 170.

16 F. Fanon, *The Wretched of the Earth*, MacGibbon & Kee (London, 1965), pp. 98-99.

17 K. Kaunda, *Humanism in Zambia and a Guide to its Implementation*, Zambia Information Services (Lusaka, 1968).

18 The Seers Report, which set the agenda for much post-colonial development policy, was particularly influential in this regard (see also Chapter Three): UN/ECA/FAO, *Economic Survey Mission on the Economic Development of Zambia* (Ndola, 1964).

19 Potts rightly argues that the level of urbanisation has been overstated, but nevertheless finds that 21% of the Zambian population was urbanised in the 1960s, higher than all other southern African countries except South Africa: D. Potts, 'Counter-Urbanization on the Zambian Copperbelt? Interpretations and Implications', *Urban Studies*, 42, 4 (2005), pp.583–609.

20 R.H. Bates, *Unions, Parties, and Political Development: a Study of Mineworkers in Zambia*, Yale University Press (New Haven CN and London, 1971).

21 B. Kapferer, *Strategy and Transaction in an African Factory*, Manchester University Press (Manchester, 1972), passim.

22 Respondent quoted in Burawoy, *The Colour of Class on the Copper Mines*, p. 80.

23 I. Henderson, 'Labour and Politics in Northern Rhodesia 1900-1953: A Study in the Limits of Colonial Power', PhD Thesis (University of Edinburgh, 1972), p. 1.

24 J. Pettman, *Zambia: Security and Conflict*, St. Martin's Press (New York, 1974), p.149.

25 G. Arrighi, and J.S. Saul, *Essays on the Political Economy of Africa*, Monthly Review Press (New York and London, 1973).

26 J. Saul, 'The Labour Aristocracy Thesis Reconsidered', in R. Sandbrook and R. Cohen (eds.) *Development of an African Working-Class: Studies in Class Formation and Action*, Longman (London, 1975).

27 See for example, 'Introduction' in Sandbrook and Cohen (eds.), *Development of an African Working-Class*; and P. Waterman, 'Aristocrats and Plebeians in African Trade Unions?: Lagos Port and Dock Worker Organisation and Struggle', PhD Thesis, Katholieke Universiteit van Nijmegen, (Den Haag, 1983); J.L. Parpart, 'The Labour Aristocracy in Africa: The Copperbelt Case 1924-1967', *African Economic History*, 13 (1984), pp. 171-191.

28 See for example, M. Legassick, 'South Africa – Forced Labour, Industrialisation and Differentiation' in R. Harris (ed.) *The Political Economy of Africa*, Schenkman (Cambridge, 1973); F.A. Johnstone, *Class, Race and Gold: a Study of Class Relations and Racial Discrimination in South Africa*, Routledge and Kegan Paul (London, 1976); H. Wolpe, 'The Theory of Internal Colonialism – The South African Case' in I. Oxaal, T. Barnett, and D. Booth (eds.) *Beyond the Sociology of Development: Economy and Society in Latin America and Africa*, Routledge & Kegan Paul (London, 1975).

29 I.R. Phimister and C. van Onselen, *Studies in the History of African Mine Labour in Colonial Zimbabwe*, Mambo Press (Gwelo, 1976).

30 A.J. Peace, *Choice, Class and Conflict: a Study of Southern Nigerian Factory Workers*, Harvester Press (Brighton, 1979), p. xi.

31 *Ibid.*
32 J. Crisp, *The Story of an African Working Class: Ghanaian Miners Struggle, 1870 – 1980*, Zed Books (London, 1984), p. 87.
33 M. Burawoy, *The Politics of Production*, Verso (London, 1985), passim.
34 Crisp, *The Story of an African Working Class*, p. 12.
35 R. Jeffries, *Class, Power and Ideology: the Railwaymen of Sekondi*, Cambridge University Press (Cambridge, 1978).
36 *Ibid.*, p. 199.
37 *Ibid.*, p. 139.
38 P.M. Lubeck, *Islam and Urban Labor in Northern Nigeria: the Making of a Muslim Working-Class*, Cambridge University Press (Cambridge, 1986, 1988), p. 3.
39 Lubeck, *Islam and Urban Labor*, p. 77.
40 Jeffries, *Class, Power and Ideology*, p. 207.
41 H. Macmillan, 'The Historiography of Transition on the Zambian Copperbelt – Another View', *Journal of Southern African Studies*, 19, 4 (1993), pp. 681-712.
42 Potts, 'Counter-Urbanization on the Zambian Copperbelt?'.
43 D. Potts, 'Shall We Go Home? Increasing Urban Poverty in African Cities and Migration Processes', *Geographical Journal*, 161, 3 (1995), pp. 245-264.
44 Ferguson, *Expectations of Modernity*, p. 124.
45 *Ibid.*, p. 19.
46 Potts notes that the World Bank's *Accelerated Development in sub-Saharan Africa: an Agenda for Action* (Washington, 1981) utilised the Zambian case study to provide evidence to support structural adjustment policies and programmes: 'Counter-Urbanization on the Zambian Copperbelt?'.
47 R.H. Bates and P. Collier, 'The Politics and Economics of Policy Reform in Zambia', in R.H. Bates and A.O. Krueger (eds.) *Political and Economic Reform: Evidence from Eight Countries*, Blackwell (Oxford & Cambridge MA, 1993).
48 E.O. Akwetey, *Trade Unions and Democratisation: A Comparative Study of Zambia and Ghana*, University of Stockholm (Stockholm, 1994).
49 *Ibid.*, p. 100.
50 *Ibid.*, p. 102.
51 L. Rakner, *Trades Unions in Processes of Democratisation: a Study of Party-Labour Relations in Zambia*, Michelsen Institute (Bergen, 1992); P. Nordlund, *Organising the Political Agora: Domination and Democratisation in Zambia and Zimbabwe*, Uppsala University (Uppsala, (1996).
52 Rakner, *Trades Unions in Processes of Democratisation*, p. 138.
53 Nordlund, *Organising the Political Agora*, p. 49.
54 *Ibid.*, p. 216.
55 T. Dunbar Moodie, with V. Ndatshe, *Going for Gold: Men, Mines and Migration*, University of California Press (Berkeley, 1994).
56 C. Brown, 'We Were All Slaves': African Miners, Culture, and Resistance at the Enugu Government Colliery, Heinemann (Portsmouth NH, 2003).
57 T.M. Klubock, Contested Communities: Class, Gender, and Politics in Chile's El Teniente Copper Mine, 1904-1951, Duke University Press (Durham NC, 1998).
58 *Ibid.*, p. 281.
59 Ibid., p. 277.

60 J.L. Parpart, Working-Class Wives and Collective Labor Action on the North Rhodesian Copperbelt, 1926-64, Pamphlet (Boston, 1985); see also H. Powdermaker, Copper Town: Changing Africa, Harper (New York, 1962).

61 A more detailed description of these source materials is provided in M. Larmer, "If We Are Still Here Next Year': Zambian Historical Research in the Context of Decline', *History in Africa* 31 (New Brunswick NJ, 2004), pp. 215-229.

62 Public meetings are described in Powdermaker, *Copper Town*, p. 116; and in Bates, *Unions, Parties, and Political Development*, pp. 106-09.

Chapter One

1 Meebelo, *African Proletarians and Colonial Capitalism*, p. 417.

2 This position builds on Cooper's arguments in *Decolonization and African Society* regarding labour in late colonial societies (see Chapter One). In its specific analysis of Northern Rhodesia, it is substantially influenced by Jane Parpart's excellent *Labour and Capital on the African Copperbelt*, Temple University Press (Philadelphia, 1983).

3 Perrings, *Black Mineworkers in Central Africa*, p. 117.

4 Henderson, 'Labour and Politics', p. 130

5 Ibid., p. 139.

6 Quoted in Berger, *Labour, Race and Colonial Rule*, p.1.

7 Quoted in Parpart, *Labour and Capital*, p. 47.

8 *Ibid.*, p. 54.

9 *Ibid.*

10 Henderson, 'Labour and Politics', pp. 125-6.

11 Parpart, Labour and Capital, pp. 31-2.

12 Quoted in Perrings, Black Mineworkers in Central Africa, p. 224.

13 The role of the RLI is detailed in L. Schumaker, *Africanizing Anthropology: Fieldwork, Networks, and the Making of Cultural Knowledge in Central Africa*, Duke University Press (Durham NC, 2001).

14 Epstein, Politics in an Urban African Community, passim.

15 Parpart, *Labour and Capital*, p. 116.

16 J. Lunn, *Capital and Labour on the Rhodesian Railway System, 1888-1947*, Macmillan (Basingstoke, 1997), pp. 135-6.

17 Cooper, *Decolonization and African Society*, passim; Henderson, 'Labour and Politics', passim.

18 Parpart, *Labour and Capital*, p. 130.

19 Henderson, 'Labour and Politics', pp. 205-6.

20 Parpart, *Labour and Capital*, p.123.

21 *Ibid.*, p. 125.

22 *Ibid.*, p. 128.

23 Henderson, 'Labour and Politics', p. 208.

24 Epstein, *Politics in an Urban African Community*, p.121.

25 Powdermaker, *Copper Town*; Parpart, *Working-Class Wives*, both passim.

26 Parpart, Working-Class Wives, p. 15.

27 quoted in Berger, Labour, Race and Colonial Rule, p. 138.

28 R.H. Bates, *Party-Union Relations in the Copperbelt of Zambia*, Yale University Press (Cambridge MA, 1969), p. 6.

29 Henderson, 'Labour and Politics', p. 230.
30 Meebelo, *African Proletarians and Colonial Capitalism*, p. 419.
31 Parpart, Labour and Capital, p.129; W.R. Jacobs, *The Relationship Between African Trade Unions and Political Organisations in Northern Rhodesia/ Zambia 1949-1961*, International Institute for Labour Studies (Geneva, 1971).
32 Parpart, *Labour and Capital*, p. 141.
33 Ibid., p. 147.
34 Jacobs, *African Trade Unions and Political Organisations*, p. 14.
35 *Ibid.*
36 Henderson, 'Labour and Politics', p. 232.
37 Quoted in Jacobs, *African Trade Unions and Political Organisations*, p. 12.
38 Parpart, *Labour and Capital*, p. 147.
39 Parpart, *Working-Class Wives*, p. 141.
40 Matthew Mwendapole, quoted in Parpart, *Labour and Capital*, p. 90.
41 Parpart, *Labour and Capital*, p. 140.
42 Meebelo, *African Proletarians and Colonial Capitalism*, p. 325.
43 Quoted in Meebelo, *African Proletarians and Colonial Capitalism*, p. 327.
44 P. Harries-Jones, *Freedom and Labour: Mobilization and Political Control on the Zambian Copperbelt*, Blackwell (Oxford, 1975), p. 98.
45 Parpart, *Labour and Capital*, p. 151.
46 Meebelo, *African Proletarians and Colonial Capitalism*, p. 429.
47 D.C. Mulford, *Zambia: The Politics of Independence 1957-1964*, Oxford University Press (Oxford, 1967), p. 341.
48 Jacobs, *African Trade Unions and Political Organisations*, p. 20.
49 Meebelo, *African Proletarians and Colonial Capitalism*, p. 439.
50 Ibid., p. 456.
51 Quoted in Jacobs, *African Trade Unions and Political Organisations*, p. 26.
52 Quoted in Meebelo, *African Proletarians and Colonial Capitalism*, p. 444.
53 Meebelo, *African Proletarians and Colonial Capitalism*, pp. 453-478.
54 *Ibid.*, p. 440.
55 *Ibid.*, p. 439.
56 *Northern News*, 14/09/62, quoted in Meebelo, *African Proletarians and Colonial Capitalism*, p. 440.
57 Meebelo, *African Proletarians and Colonial Capitalism*, p. 461.
58 Meebelo describes the deposit of ICFTU funds into personal bank accounts, and widespread corruption in the union movement: *African Proletarians and Colonial Capitalism*, p. 457.
59 *Ibid*, p. 468.
60 *Ibid.*, p. 506.
61 Quoted in C. Gertzel, 'Industrial Relations in Zambia to 1975', in U.G. Damachi, H.D. Seibel, and L. Trachtman (eds.) *Industrial Relations in Africa*, St. Martin's Press (New York, 1979), p. 320.
62 *Voice of UNIP*, May-June 1964, quoted in Meebelo, *African Proletarians and Colonial Capitalism*, p. 487.
63 R. Hall, *Zambia, 1890-1964: The Colonial Period*, Longman (London 1965, 1976), p. 118.

Chapter Two

1 Zambia, *Towards Complete Independence*, Government Publishers (Lusaka, 1969).

2 United Nations/Economic Commission for Africa /Food & Agriculture Organisation, *Economic Survey Mission on the Economic Development of Zambia*, (Ndola, 1964), passim.

3 *Ibid.*, p. 27.

4 *Ibid.*, p. 33.

5 *Northern News*, 2/12/64.

6 A. Roberts, *A History of Zambia*, Heinemann (London, 1976), pp. 229-230.

7 *Ibid.*, p. 227.

8 C. Baylies, 'State and Class in Postcolonial Africa' in M. Zeitlin, *Political Power and Social Theory Vol 5*, JAI Press (1985), pp. 1-34.

9 Roberts, *A History of Zambia*, p. 229.

10 See for example, Tordoff, *Politics in Zambia*; B. Turok, (ed.), *Development in Zambia – A Reader*, Zed Books (London, 1979).

11 See in particular G. Macola, '"It Means as if We Are Excluded from the Good Freedom": Thwarted Expectations of Independence and Royal Politics in the Luapula Province of Zambia, 1964-1966', *Journal of African History*, 47, 1 (2006); and M. Larmer, "A little bit like a volcano' – The United Progressive Party and resistance to one-party rule in Zambia, 1964 – 1980', *International Journal of African Historical Studies*, forthcoming, 2006.

12 C. Gertzel, (ed.), C. Baylies, and M. Szeftel, *The Dynamics of the One-Party State in Zambia*, Manchester University Press (Manchester, 1984), passim.

13 *Ibid.*, p. 19.

14 *Times of Zambia* (hereafter *ToZ*), 16/8/76.

15 *Ibid.*, 17/8/73.

16 Andrew Sardanis argues that Lonrho's Tiny Rowland's personal links to Kaunda enabled him to earn substantial profits from the trading of ZIMCO bonds: *Africa: Another Side of the Coin*, I B Tauris (New York, 2003), pp. 266-271.

17 Liwewe, PR Lusaka, to Musukuma, *Mining Mirror*, 14/10/74, Zambia Consolidated Copper Mines archives (hereafter ZCCM) 15.2.B, 'Ministry of Mines, annual reports and correspondence, 1972-78'.

18 *ToZ*, 25/6/79.

19 *Ibid.*, 1/6/79.

20 M. Szeftel, 'Political Graft and the Spoils System in Zambia - the State as a Resource in Itself', *Review of African Political Economy* 24 (1983), passim.

21 *ToZ*, 12/5/77.

22 J. Mwanakatwe, *Teacher, Politician, Lawyer: My Autobiography*, Bookworld Publishers (Lusaka, 2003), pp. 282-3; Gertzel, 'Dissent and Authority in the Zambian One-Party State', in *The Dynamics of the One-Party State*, p.88.

23 NCCM Annual Report 1978, ZCCM/10.2.9B, 'NCCM Annual Reports, 1971-81'.

24 *ToZ*, 18/3/78.

25 *Zambia Daily Mail*, 15/2/79.

26 Kenneth Kaunda address to leaders, employers and workers, Mulungushi Hall. 26/6/80, in Mineworkers' Union of Zambia Head Office (hereafter MUZ HO), 'ZCTU-related papers, 1977-92'.

27 See for example the recommendations in UNIP archives (hereafter UNIP) 8.8.4, H.K. Maunga memo, 'General and Presidential Elections 1978', 24/7/78; Larmer, 'A Little Bit Like a Volcano', p.38.
28 Larmer, 'A Little Bit Like a Volcano', p.18.
29 F. Bridgland, *Jonas Savimbi: a Key to Africa*, Mainstream Publishing (Edinburgh, 1986), pp. 155-178.
30 *ToZ*, 23/11/79.
31 See for example article in *Sunday Times of Zambia* (hereafter *SToZ*), 6/7/80.
32 C. Gertzel, 'Dissent and Authority in the Zambian One-Party State 1973-80', p. 91.
33 *ToZ*, 5/2/74.
34 *Ibid.*, 1/10/76.
35 Amongst many examples, see UNIP 8.9.11, Kasama tour report, Alex Shapi and Frank Chitambala, 28/8-21/9/74; 8.9.11, Registration of Voters report, Kaputa, Luwingu, Mporokoso, 5-23/4/75, J. Mutala [Mutale?]; 8.4.4, Copperbelt Province report to National Council, 14-18/12/74 written by Minister for Copperbelt A.C. Chalikulima, 13/12/74; 8.4.3, Minister of Lands S. Kalulu, tour of his Lusaka constituency, 12/4/73.
36 Interviews with former UPP leaders and activists: John Chisata, 26-27/11/2003; Eric Bwalya, 1/12/2003; Faustino Lombe, 11/12/2003.
37 *ToZ*, 23/12/77.
38 *Ibid.*, 30/3/78.
39 *Ibid.*, 8/6/78.
40 UNIP 8.8.4, Member of Central Committee J. Litana, to Election and Political Strategy sub-Committee, recommendations to Central Committee, n.d., c. start 1979.
41 *Munyama Human Rights Commission Report* (Lusaka, 1995).
42 Article, Akashambatwa Lewanika, in *SToZ*, 5/7/87.
43 *ToZ*, 24-28/7/84.
44 *Ibid.*, 12/5/90.
45 Reported in *ToZ*, 26/9/86.
46 Potts, 'Counter-Urbanization on the Zambian Copperbelt?', passim.
47 *SToZ*, 30/5/82.
48 *Ibid.*, 1/9/85.
49 *ToZ*, 27/11/85.
50 Ibid., 3/8/83.
51 Sardanis, Africa: Another Side of the Coin, p. 292.
52 See for example, letters in *SToZ*, 1/1/84.
53 *ToZ*, 30/4/82.
54 J. Aron, 'Economic Policy in a Mineral-Dependent Economy: the Case of Zambia', DPhil Thesis, University of Oxford (1992), pp. 202-205.
55 *Zambia Mining Year Book (Special Edition), 1982-86*, ZCCM (Lusaka, 1987).
56 ZCCM Annual Report 1985, ZCCM/10.2.8A, 'ZCCM Annual Reports, 1985 – 1992'.
57 Ibid.
58 Aron, 'Economic Policy in a Mineral-Dependent Economy', p. 205.
59 *ToZ*, 21/2/87.

60 ZCCM Management Brief, 30/1/86, ZCCM/16.2.9C, 'GIRM Committee and related correspondence, 1986'.
61 *ToZ*, 23/5/84.
62 *Ibid.*, 15/10/83.
63 *Ibid.*, 2/2/85 & 22/5/85.
64 *Ibid.*, 5/10/85.
65 *Ibid.*, 12/10/85.
66 *SToZ*, 20/10/85.
67 *ToZ*, 16/12/85.
68 E. West, 'The Politics of Hope: Zambia's Structural Adjustment Programme 1985-1987', PhD Thesis, Yale University (1989), p. 224; Bates and Collier, 'The Politics and Economics of Policy Reform', p. 412.
69 *ToZ*, 1/4/86; *SToZ*, 26/10/86.
70 *ToZ*, 30/6/87.
71 *Ibid.*, 5/3/87.
72 *Ibid.*, 4/4/87. Nordlund argues whilst these strikes were of 'economic origin', their impact was politically significant: *Organising the Political Agora*, p. 91.
73 *Ibid.*, 9/4/87.
74 World Bank Country Economics Department WPS 222/C. Colclough, *The Labour Market and Economic Stabilisation in Zambia*, World Bank (Washington, 1989), p. 1.
75 *ToZ*, 10/12/87.
76 *Ibid.*, 7/11/88, 13/5/89.
77 *Ibid.*, 10/11/88.
78 *SToZ*, 19/11/89.
79 *ToZ*, 3/7/89.
80 *Ibid.*, 1/7/89, 17/7/89.
81 *Ibid.*, 10/7/89.
82 *Ibid.*, 12/10/89.
83 *Ibid.*, 15/11/88.
84 *Ibid.*, 1/1/90.
85 *Ibid.*, 9/4/90.
86 *Ibid.*, 6/2/90.
87 *Ibid.*, 25/3/82.
88 *Ibid.*, 8/5/82.
89 *Ibid.*, 5/10/87.
90 *Ibid.*, 21/3/85.
91 *Ibid.*, 16/4/88; 4/5/88.
92 *Ibid.*, 2/12/84.
93 *SToZ*, 3/1/82.
94 *Ibid.*, 11/2/83.
95 For example Andrew Kashita, in *ToZ*, 12/6/89.
96 *ToZ*, 6/8/82.
97 Nordlund, *Organising the Political Agora*, passim.
98 *ToZ*, 6/2/90.
99 *Ibid.*, 16/3/89.
100 *Ibid.*, 20/7/88, 22-23/8/88.
101 *SToZ*, 31/12/89; Nordlund, *Organising the Political Agora*, p. 93.

102 *SToZ*, 13/5/90; *ToZ*, 14/5/90.
103 *ToZ*, 29/5/90.
104 *Ibid.*, 20/6/90.
105 *Ibid.*, 26/6/90.
106 Nordlund, *Organising the Political Agora*, p. 101.
107 *ToZ*, 21/7/90.
108 *Ibid.*, 6/8/90.
109 *Ibid.*, 26/7/90, 25/9/90.
110 *Ibid.*, 18/10/90.
111 *Ibid.*, 3/6/91.
112 Nordlund, *Organising the Political Agora*, p. 106.
113 *ToZ*, 21/12/90, 10/1/91.
114 *Ibid.*, 2/3/91.
115 Interview with Arthur Wina, July 1991.
116 *ToZ*, 21/11/90.
117 Rakner, *Political and Economic Liberalisation in Zambia*, p. 68; *ToZ*, 14/2/91.
118 *SToZ*, 26/5/91.
119 *ToZ*, 24/6/91.
120 Member of UNIP Central Committee Joseph Mutale, quoted in *ToZ*, 15/2/91.
121 *National Mirror*, 1/7/91.
122 *ToZ*, 6/11/91.
123 *Ibid.*, 5/11/91.

Chapter Three

1 Social conflict in post-colonial Zambia was by no means limited to organised labour, as Giacomo Macola eloquently demonstrates with regard to regional/chiefly politics: "'It Means as if We Are Excluded from the Good Freedom'".
2 P. J. Schraeder, African Politics and Society: a Mosaic in Transformations Bedford/St. Martins (Boston & New York, 2000), p. 218.
3 R.H. Bates, Unions, Parties, and Political Development: a Study of Mineworkers in Zambia, Yale University Press (New Haven CN & London, 1971), p. 1.
4 See Gertzel's review of Bates and Burawoy: C. Gertzel, 'Labour and the State: The Case of Zambia's Mineworkers' Union', *Journal of Commonwealth and Comparative Politics* 13, 3 (1975).
5 Cooper, *Decolonization and African Society*, p. 7.
6 W. Tordoff, *Government and Politics in Africa* 4th Edition, Palgrave-MacMillan (Basingstoke, 2002), p. 92.
7 Nordlund, *Organising the Political Agora*, p. 61.
8 Burawoy, *The Colour of Class on the Copper Mines*, p. 90.
9 *Ibid.*, p. 2.
10 *Ibid.*, p. 77.
11 *ToZ*, 16/4/65.
12 *Northern News*, 16/11/64.
13 *Ibid.*, 17/12/64.

14 Cabinet Legislation Committee undated, c. October 1964, National Archives of Zambia (hereafter NAZ)/LSS1/20/89, 'Trade Unions & Trades Disputes amendment, 1964-66'.
15 B.B. Liatto, 'Organised Labour and the State in Zambia', PhD Thesis, University of Leeds (1989).
16 *Northern News*, 8/6/64.
17 MoL Asst Secretary M.H.M. Shalo, Record Note, UTUC Executive Board meeting, 2/10/65, item 33, NAZ/LSS1/19/226, 'UTUC soon to be ZCTU, 1965-66'.
18 *ToZ*, 27/10/65.
19 Minutes of last Executive Board meeting, UTUC, 31/10/65, NAZ/LSS1/19/220, 'UTUC, 1964-65'.
20 *ToZ*, 1/12/65.
21 *Daily Hansard* Columns 135-145, 14/2/65, item 118, NAZ/LSS1/19/226, 'UTUC soon to be ZCTU, 1965-66'.
22 *ToZ*, 17/12/66.
23 General Secretary Nkumbula, 26/1/66, item 48, NAZ/LSS1/19/227, 'ZCTU, 1965-67'.
24 Acting Asst Secretary to Under Secretary, 29/6/66, NAZ/LSS1/19/226, 'UTUC soon to be ZCTU, 1965-66'; see for example General Secretary to PS Ministry of Labour, 9/7/66, NAZ/LSS1/19/227, 'ZCTU, 1965-67'.
25 Registrar of Trade Unions [Nyondo] to Permanent Secretary Kasonde, 11/10/66, item 249, NAZ/LSS1/19/238, 'ZCTU, 1966-68'.
26 *ToZ*, 12/12/67.
27 GS ZCTU to member unions, 9/7/69, item 77, and unnumbered document, both NAZ/LSS1/19/260, 'ZCTU, 1969'.
28 S.K. Makuka for acting GS ZCTU to Labour Commissioner, 10/6/71, NAZ/LSS1/20/96, 'ZCTU, 1971'.
29 Various correspondence and meetings, NAZ/LSS1/19/243, 'ZCTU, 1966-68'.
30 RTU, n.d., item 342, NAZ/LSS1/19/238, 'ZCTU, 1966-68'.
31 *ToZ*, 10/12/68.
32 *Ibid.*, 28/4/70.
33 *Ibid.*, 29/2/68, 14/8/68, 26/8/68.
34 *Ibid.*, 27/12/68, 28/12/69, 9/1/69.
35 *Ibid.*, 9/5/70.
36 October 1964 Report, NAZ/LSS1/15/297, 'Monthly reports – Ndola'.
37 1st General Managers meetings, July and December 1964, ZCCM/1.7.5C, 'African Labour Political 1964-65'.
38 November 1964 Report, NAZ/LSS1/15/297, 'Monthly reports – Ndola'.
39 CISB to companies, 28/12/64, ZCCM/17.2.3B, 'ZMU local disputes NCR, 1964-67'.
40 February 1965 Report, NAZ/LSS1/15/297, 'Monthly reports – Ndola'.
41 NCR (Wright) to Companies (Reid), 5/3/65, ZCCM/11.6.2A, 'Conciliations, arbitrations, strikes, disputes, 1965'.
42 *ToZ*, 19/3/65.
43 *Ibid.*, 22/3/65.
44 *Ibid.*

45 Dept of Labour Monthly Report March 1965, item 16, NAZ/LSS1/15/277, 'Dept of Labour monthly report, 1964-66'.
46 CISB Industrial Relations report March 1965, ZCCM/10.6.10F, 'Monthly comments on Industrial Relations'.
47 Chingola to CISB, 24/3/65, ZCCM/17.2.3B, 'ZMU local disputes NCR, 1964-67'. Such reports demonstrate that mine management had inside information of ZMU Supreme Council meetings.
48 Ibid.
49 CISB to Companies, 14/4/65, ZCCM/17.2.3B, 'ZMU local disputes NCR, 1964-67'.
50 CISB to Companies, 21/4/65, in Ibid.
51 Dept of Labour Monthly Report May 1965, item 18, NAZ/LSS1/15/277, 'Dept of Labour monthly report, 1964-66'; 3rd MJIC meeting, 29/10/65, ZCCM/16.3.2B, 'MJIC-MUZ negotiations, 1965-66'.
52 MJIC sub-committee meeting, 22/1/65, ZCCM/16.3.2B, 'MJIC-MUZ negotiations, 1965-66'.
53 ZMTU [ZMU] Memo on Wage Dispute with the Mining Companies, 27/3/65, ZCCM/11.6.2A, 'Conciliations, arbitrations, strikes, disputes, 1965'.
54 Sam Kangwa interview, 5/10/2002.
55 AAC (Gavin Relly) and RST (James Reid) to Minister of Labour, 28/9/65, NAZ/LSS1/1.9.2F, 'ZMU General, May – Nov 1965'.
56 ZMU Emergency Supreme Council, 27/2/65, in MUZ Luanshya, 'Supreme Council 1964-1982'.
57 Notes on conciliation proceedings, 5-6/6/65; CISB to Companies, conciliation proceedings, 11/6/65; both in ZCCM/11.6.2A, 'Conciliations, arbitrations, strikes, disputes, 1965'.
58 ZMU Supreme Council, 31/7 – 1/8/65, in MUZ Luanshya, 'Supreme Council 1964-1982'.
59 *ToZ*, 11/6/65; MoL Mundia to PS, 21/4/65, item 342, NAZ LSS1/19/224, 'NRh AMTWU, 1964-66'.
60 Abel Musonda interview, 20/3/2003.
61 ZMU GS Mweene to RTU, 28/9/65, item 19, NAZ/LSS1/19/228, 'ZMU, 1965-67'.
62 Record note by Acting RTU, 12/10/65, item 23, Ibid.
63 CISB to Companies, 16/10/65, ZCCM 1.9.2F, 'ZMU General, May-Nov 1965'.
64 Bates, *Unions, Parties, and Political Development*, p. 155.
65 Emergency Supreme Council 9/12/65, in MUZ Luanshya, 'Supreme Council 1964-1982'.
66 ZMU Emergency SC, 28/12/65, in MUZ Nkana, 'MUZ SC and CEC mtgs, other HO, 1970-1987'.
67 '*Kaspato*' leaflet, undated, item 75, NAZ/LSS1/19/228, 'ZMU, 1965-67'. *Kasapato* was a column in UNIP publications.
68 LO P.K. Musulwa report, 18/12/65, item 45/1, NAZ/LSS1/19/228, 'ZMU, 1965-67'.
69 ALC Sabel to Acting Under-Secretary, 24/12/65, item 600, NAZ/LSS1/19/224, 'NRh AMTWU, 1964-66'.
70 *ToZ*, 5/2/66.

71 Bates, *Unions, Parties, and Political Development*, p. 165.
72 *ToZ*, 20/1/66.
73 ZMU GS Cosmas Mweene (reported in ToZ, 5/4/66): 'It was a time of severe stress caused by UDI. We felt we had a duty not to prolong our demands.'
74 CISB to Companies, General Managers, 30/3/66, ZCCM/16.2.8F, 'Work Stoppage, ZMU Local Disputes Bancroft, 1965-67'.
75 *ToZ*, 2/4/66.
76 *Ibid.*, 6/4/66.
77 Chibuluma/Kalulushi to NCR, 6/4/66, ZCCM/17.2.2C, 'Local disputes, work stoppages, Chibuluma, 1965-68'.
78 Reported in *ToZ*, 8/4/66.
79 Articles by Martin Meredith, *ToZ*, 8/2/66 and 17/4/66.
80 Powdermaker, *Copper Town*, p. 116; Bates, *Unions, Parties, and Political Development*, pp. 106-109.
81 Frank Juretic, undated (c.May 1966), 'Notes on Mr.Leslie's comments', 1.9.2F, 'ZMU General, May-Sept 1966'.
82 *ToZ*, 29/6/66.
83 Sikota Wina, 'Report of a mtg held by Messrs' Sikota Wina and Milner and Representatives of ZMU, MLSA and MAPA in the Chawama Hall' [Mufulira], 28/8/66, ZCCM/1.9.2F, 'ZMU General, May – Sept 1966'.
84 *ToZ* editorial, 31/8/66: 'When the rebel element in such a huge trade union – the biggest in independent Africa – can defy its leaders with impunity and encourage the rank and file to follow suit, it is time for stern action.'
85 *ToZ*, 30/8/66.
86 NCR Strike Diary, 30/8/66 – 21/9/66, ZCCM/1.9.2F, 'ZMU General, May-Sept 1966'.
87 Ibid.
88 *ToZ*, 15/9/66.
89 *Ibid.*, 21/9/66.
90 CISB to General Managers, 14/9/66, ZCCM/11.6.2A, 'Conciliations, arbitrations, strikes, disputes, 1966-68'.
91 Fidelis Mwamba interview, 15/9/2002.
92 Ibid.
93 *ToZ*, 21/5/66.
94 *Ibid.*, 8/6/66.
95 Bates, *Unions, Parties, and Political Development*, pp. 40-44.
96 Liatto, 'Organised Labour and the State', p. 43.
97 Timothy Walamba interview, 6/11/2002.
98 For example, Basilio Kunda interview, 7/2/2003.
99 Walamba interview.
100 Mwamba interview.
101 RTU meeting with ZMU officials, 5/9/66, item 103, NAZ/LSS1/19/227, 'ZCTU, 1965-67'.
102 *ToZ*, 29/3/67, argued that a ballot of members of the three unions was unlikely to deliver a vote in favour.
103 CISB to mine companies, 5/5/67, ZCCM/19.2.11D, 'Meetings, disputes and union affairs, 1967'.

104 *The Miner*, 2/6/67, item 24, NAZ/LSS1/19/245, 'MUZ, 1967-69'.
105 Gertzel, 'Labour and the State', p. 300.
106 This was a substantial strengthening of the Reformed Trade Union Congress position, adopted in 1960 and subsequently informing the ZCTU's position, of supporting 'the most progressive party'. This was applied to UNIP, but was implicitly transferable to different parties. General Secretary Thawe's report to MUZ Conference 1968, MUZ HO, 'Annual Conferences, MUZ governing bodies, 1967-70'.
107 Ibid.
108 Presidential Speech to Quarterly SC, 14/1/68, MUZ Luanshya/5A 'Supreme Council, 1967-68'.
109 MJIC meeting, 23/11/66, ZCCM/17.2.3B, 'ZMU general, 1966'.
110 *ToZ*, 25/4/68.
111 B. Kapferer, *Strategy and Transaction in an African Factory*, Manchester University Press Imprint (Manchester, 1972), passim.
112 Gertzel, 'Industrial Relations in Zambia', p. 345.
113 Zambia Ministry of Labour, *Livingstone Labour Conference, 7th-8th April 1967*, Government Printers (Lusaka, 1967), p. 8.
114 *Ibid.*, p. 32.
115 International Labour Organisation/United Nations Development Programme/Technical Advisory Service, *Report to the Government of Zambia on Incomes, Wages & Prices in Zambia: Policy & Machinery* (Geneva, 1969).
116 James Fry article in *ToZ*, 3/4/70.
117 Turner, 1969, quoted in Burawoy, *The Colour of Class on the Copper Mines*, p. 56.
118 M. Hamalengwa, *Class Struggles in Zambia, 1889-1989, and the Fall of Kenneth Kaunda, 1990-1991*, University Press of America (Lanham MD and London, 1992), pp. 55-56.
119 'Towards a new concept of trade unionism', *ToZ*, 17/4/69.
120 *Ibid.*
121 1st Mtg of Ctte on Zambianisation in the Mining Industry, 20/1/67, ZCCM/19.3.2C, 'Konkola, work stoppages, strikes, man hours lost, 1966-67'.
122 Ibid.
123 2nd MJIC meeting, 26/6/68, in MUZ Luanshya, 'BVC H Chibwe's file, 1963-69'.
124 M. Burawoy, 'The Hidden Abode of Under-Development: Labour Process and the State in Zambia', *Politics and Society*, 11, 2 (1982), p. 155.
125 Burawoy, *The Colour of Class on the Copper Mines*, passim.
126 *ToZ*, 18/10/68.
127 3rd MJIC meeting, 27/9/68, MUZ Luanshya/5/D, 'CISB-MJIC mtgs, 1967-69'.
128 *ToZ*, 21/3/69, 24/3/69; *Financial Times*, 26/3/69.
129 *ToZ*, 31/3/69, 4/4/69.
130 Companies' 'Statement of Case' to Conciliation Board, 3/4/69, ZCCM/11.6.2A, 'Conciliations, arbitrations, strikes, disputes, 1968-70'.
131 *ToZ*, 16/5/69.
132 *Ibid.*, 14/11/69.
133 *Zambia Daily Mail*, 31/3/70, ZCCM/3.1.4F, 'MUZ elections etc, 1970'.
134 MJIC sub-committee meeting, Ibid.

135 Copper Mining Companies and MUZ, 'Agreement for a Package Deal Incorporating an Integrated Pay Structure, Improvements to Conditions of Service and Measures to Increase Efficiency', 24/7/70, in MUZ HO, 'MJIC meetings, 1970'.
136 *ToZ*, 22/9/70.
137 Letter, Henry Kabache, Luanshya in *ToZ*, 29/3/71.
138 *ToZ*, 9/9/70.
139 Zambia, *Towards Complete Independence*, Government Printers (Lusaka, 1969).
140 *Ibid.*, p. 44.
141 *Ibid.*, p. 54.
142 Bates, *Unions, Parties, and Political Development*, p. 213.
143 Burawoy, *The Colour of Class on the Copper Mines*, p. 90.
144 *Northern News*, 12/12/64.
145 Quoted in F.O.E. Okafor, 'The Mining Multinationals and the Zambian Economy', *The African Review*, 17, 1 & 2 (1990), p. 74.
146 McCourt, Mufulira to CISB Kitwe, 4/9/69, ZCCM/15.2.1C, 'Government Takeover, 1969 – 1971'.
147 Discussions at State House on 'Co-determination', 13/11/69, item 6, NAZ/LSS1/17/70.
148 Kaunda to Mulikita, Re: 'Layman's Draft', 3/8/70, item 45, Ibid.
149 'Report on Workers' Seminar', Mufulira, 20-22/3/70, item 33, NAZ/LSS1/17/70.
150 John Price, Executive Officer ZFE to MoL Mulikita, 11/3/70, item 28, NAZ/LSS1/17/70.
151 MoL Chakulya to Mindeco Managing Director Dominic Mulaisho, 31/12/71, ZCCM/1.5.3G, 'Industrial Relations Act 1971, 1970-1995'.
152 CISB Personnel Research Unit, R.J. Wenham, 'Worker Participation', 15/2/72, ZCCM/1.9.1D, 'IRB, ACME-MJIC mtgs, Feb 1982 – May 1987'.
153 See for example, *ToZ*, 18/10/68.
154 *ToZ*, 21-27/12/70.
155 *Ibid.*, 19/2/70.
156 *Ibid.*
157 *Ibid.*, 11/3/70.
158 *Ibid.*, 11/7/70.
159 'Branch Officials elected 20/10/70 and 30/10/70', ZCCM/3.1.4F, 'MUZ – General, 1967-1970'.
160 *ToZ*, 27/10/70.
161 *Ibid.*, 31/10/70 and 4/11/70.
162 *Ibid.*, 16/11/70.
163 Abel Musonda interview, 20/3/2003.
164 *ToZ*, 15/3/71.
165 *Ibid.*, 18/3/71.
166 *Ibid.*, 29/3/71.
167 Roy Yumbe interview, 7/11/2002.
168 *Zambia Daily Mail*, 21/1/71.
169 *ToZ*, 6/4/71.
170 *Ibid.*, 9-12/4/71 and 19/4/71.
171 *Ibid.*, 21/4/71 and 24/4/71.

172 *Ibid.*, 3/5/71.
173 Item 53, NAZ LSS1/19/269, 'MUZ, 1971'.
174 Ibid.
175 *ToZ*, 12/5/71.
176 *Ibid.*, 13/3/71.
177 CISB to companies, 13/5/71, ZCCM/7.1.4I, 'IRB MJIC, MUZ mtgs, mins etc, 1970-74'.
178 *ToZ*, 21/5/71.
179 *Ibid.*, 26/5/71.
180 *Ibid.*, 31/5 –2/6/71.
181 Gertzel, 'Labour and the State', p. 293.
182 Burawoy, *The Colour of Class on the Copper Mines*, p. 48.
183 Gertzel, 'Labour and the State', p. 295.
184 R.H. Bates, *Rural Responses to Industrialization: a study of village Zambia*, Yale University Press (New Haven & London, 1976), p.242.
185 *Zambia Daily Mail*, 2/8/71, cited in Bates, *Rural Responses to Industrialization*, p.332.
186 MUZ Supreme Council minutes, 24/9/71, MUZ Luanshya/5B, 'Supreme Council, 1971-73'.
187 Chisata interview.
188 Ibid.
189 G.B. Mwangilwa, *The Kapwepwe Diaries*, Multimedia Publications (Lusaka, 1986), p. 145.
190 Mwamba interview.
191 Basilio Kunda interview, 7/2/2003.
192 Chisata interview; Sefelino Mumba interview.
193 *ToZ*, 22/12/71.
194 Mwangilwa cites a Daily Mail interview with a miner who had refused to vote in protest to UNIP membership card-checking campaigns in the constituency: *The Kapwepwe Diaries*, p. 151.
195 MUZ Head Office (Confidential), 17/1/72: 'Summary of circumstances which led to the loss of the Mufulira West constituency seat in the Parliamentary by-elections by a UNIP candidate – Comrade Alexander Kamalondo', in papers of Peter Matoka, NAZ/HM77/UNIP/1/4. Un-authored, but apparently written by MUZ Organising Secretaries Peter Kabungo and Morrison Sumani, sent to Mufulira to help the UNIP campaign. Submitted to the MUZ Supreme Council 18/1/72, minutes in MUZ Luanshya/5B, 'Supreme Council, 1970-73'.
196 Ibid.
197 Ibid.
198 Ibid.
199 MUZ Supreme Council, 18/1/72, in Ibid.
200 MUZ Supreme Council, 9/6/72, in Ibid.
201 UNIP Luanshya West Constituency Vice Secretary to Secretary MUZ Roan, 27/4/72, in MUZ Luanshya/17A, 'UNIP, 1971-74'; *ToZ*, 22/5/72.
202 Luka Lukulu, at same meeting, Ibid.
203 *ToZ*, 31/5/72.

204 Acting Labour Commissioner Sketchley Sacika to Shaw & Co, Chingola, 17/6/71, item 646; Acting LC Sacika to Acting PS, 17/6/71, item 649; both NAZ/LSS1/20/96, 'ZCTU, 1971'.
205 Minister Chakulya's speech to ZCTU General Council, 10/7/71, item 665, Ibid.
206 ALC, Record Note of General Council, 11/7/71, item 687, Ibid.
207 Quoted in *ToZ*, 27/4/72.
208 *ToZ*, 5/9/72.

Chapter Four

1 C. Baylies and M. Szeftel, 'Elections in the One-Party State', in Gertzel (ed.), *The Dynamics of the One-Party State.*
2 Gertzel, 'Labour and the State', p. 295.
3 International Marxist Group, *Zambia: Humanist Rhetoric, Capitalist Reality*, Africa in Struggle Occasional Papers No.2 (London, 1975), p. 21.
4 See Phimister and van Onselen, *Studies in the History of African Mine Labour*, passim.
5 ZCTU, resolutions from Seminar of GSs and Presidents held at Serenje, 12-17/3/73, in MUZ Nkana, 'ZCTU and Labour Day'.
6 *SToZ*, 25/2/73.
7 *Ibid.*, 8/2/76.
8 Akwetey, *Trade Unions and Democratisation*, p. 50.
9 *ToZ*, 20/11/74.
10 *Ibid.*, 26/11/74.
11 Timothy Walamba interview, 6/11/2002.
12 *ToZ*, 5-6/3/78.
13 Gertzel, 'Dissent and Authority in the Zambian One-Party State', p. 92.
14 *ToZ*, 26-27/7/74.
15 *Ibid.*, 11/8/75.
16 Akwetey, *Trade Unions and Democratisation*, p. 48.
17 *Ibid.*, p. 62.
18 N. Zimba, ZCTU SG, address to 1974 MUZ conference, in MUZ HO, 'annual conferences, government bodies'.
19 'Statement on Employment, Income Distribution and Social Progress' at Symposium, 26-31/1/76, in MUZ Nchanga, miscellaneous documents.
20 UNIP Secretary General Circular 11 of 1974, quoted in RCM Mufulira, summary of guidelines on formation of Political Committees at Mufulira Division, ZCCM/15.2.1C, 'Government, general political, 1973-78'.
21 Steven Mukuka interview, 18/3/2003.
22 Liatto, 'Organised Labour and the State in Zambia', p. 22.
23 Gertzel, 'Dissent and Authority in the Zambian One-Party State 1973-80', p. 92.
24 ZCTU 'Statement on Multinational Companies', Symposium, 26-31/1/76, in MUZ Nchanga, miscellaneous documents.
25 *ToZ*, 3/5/77.
26 *SToZ*, 17/8/77.
27 ZCTU, 'Observations and Questions on Trends and Development in the Zambian Economy', 1/11/78, in MUZ Nkana, 'ZCTU and Labour Day'.

28 *ToZ*, 31/7/78.
29 *Ibid.*, 26/8/79.
30 *Ibid.*, 6/1/79.
31 Article, *ToZ*, 10/4/76.
32 *ToZ*, 18/4/76; Hamalengwa, *Class Struggles in Zambia*, p. 81.
33 *ToZ*, 6/10/77. Union Presidents were known as 'National Chairmen' during the 1970s and 1980s, because only the state President was permitted to use the title.
34 See for example, 'Luanshya Labour Office Annual Report, 1978', NAZ/LSS/DL/102/1/17, 'Luanshya Annual Reports, 1969 – 1995'.
35 CISB telex, 'Court Case, WC Luanshya vs Luanshya Management, RCM Ltd', 20/10/78, ZCCM/1.5.3G, 'Industrial Relations Act, 1970-1995'.
36 R. Fincham, and G. Zulu, 'Labour and Participation in Zambia', in Turok (ed.) *Development in Zambia – A Reader.*
37 *ToZ*, 16/11/76.
38 Submission by D. Mwila, 'Workers' Participation in Equity', July – Dec 1977, in MUZ HO, IB/B/71, 'Other Govt Ministries and Govt Bodies, 1968-1981'.
39 *ToZ*, 14/11/77.
40 CIRS-CISB memo to RCM/NCCM, 'Changes to Constitution', 10/1/73, ZCCM/1.4.5B, 'MUZ Constitution, 1967-95'.
41 CEC minutes, 5/3/75, in MUZ HO, 'Supreme Council and CEC minutes 1975-1980'.
42 For example David Ndalama interview, 13/2/2003.
43 For example interviews with Charity M'tonga, 30/1/2003; Alex Kalunga, 23/1/2003.
44 Fenwick Nyirongo interview, 10/2/2003; Pascale Mwansa interview, 26/2/2003.
45 'Crucified Miner', letter to *ToZ*, 21/5/73.
46 Burawoy, *The Colour of Class on the Copper Mines*, passim.
47 'Frustrated Miner', letter to *ToZ*, 12/4/73.
48 *ToZ*, 8/1/74.
49 Quoted in *ToZ*, 16/4/74.
50 *ToZ*, 27/5/74.
51 *Ibid.*, 20/5/74; Yumbe was still apparently suspended from MUZ following his detention: MUZ SC minutes, 28/5/74, in MUZ HO/5/8/73 'Supreme Council Minutes, 1972-76'.
52 Joint Executive Committee, 6/1/76, in MUZ Luanshya 6A, 'Luanshya JEC and caucus meetings, 1979-81'.
53 *ToZ*, 18/7/74; see also Hamalengwa, *Class Struggles in Zambia*, p. 58.
54 *Ibid.*, 28/8/74.
55 *Ibid.*, 1/9/76.
56 Quoted in *ToZ*, 26/7/74.
57 n.d., c. end July 1974, Shapi to Kaunda, in papers of Alex Shapi, NAZ HM/89.
58 Mwila speech to Biennial Conference 1974, in MUZ HO, 'Annual conference, governing bodies'

59 MUZ, Resolutions for Biennial Conference, Livingstone, 28-30/8/74, ZCCM/3.1.4F, 'MUZ, 1974'.
60 I am grateful to Mike Murphy for making this point to me.
61 Fidelis Mwamba interview, 15/9/2002.
62 See for example, Ndola MUZ branch submission to 1976 Conference, in MUZ HO, 'Annual conference files'.
63 Ibid; Robert Kapenda interview, 22 January 2003.
64 Maxwell Kalesha-Chisoko interview, 6-7 February 2003.
65 Reported in *ToZ*, 18/10/76.
66 Unnamed employees in Stores Dept to Head of Dept, n.d., March 1976, in MUZ Luanshya untitled file containing grievance cases correspondence, 1970-81.
67 A.J. Soko, Briefing to NCCM and RCM GMs, Mufulira, 20/10/76, re improved Expatriate Conditions of Service, ZCCM/15.2.6B, 'Ministry of Mines, annual reports & correspondence, 1972-78'.
68 Daniel, *Africanisation, Nationalisation and Inequality*, pp. 151-2.
69 Joint Caucus meeting, 13/9/78, in MUZ Luanshya/6A, 'Luanshya JEC and caucus mtgs, 1979-81'; *ToZ*, 30-31/8/75.
70 *Mining Mirror*, 5/9/75, ZCCM/16.2.9B, 'MJIC-MUZ-ACME mtgs, 1974-75'.
71 Reported in *ToZ*, 5/10/76; also *ToZ*, 9/10/76.
72 N.N.J. Mijere, 'The Mineworkers' Resistance to Governmental Decentralisation in Zambia: Nation-Building and Labor Aristocracy in the Third World', PhD Thesis, Brandeis University (1985), p. 317.
73 Basilio Kunda interview, 7/2/2003.
74 Manjolo Banda interview, 14/2/2003.
75 Roy Yumbe interview, 7/11/2002.
76 For example George Lombe interview, 27/2/2003.
77 Stephen Musonda interview, 29/1/2003.
78 For example Raphael Kasonde interview, 23/2/2003.
79 4th MJIC, 21/9/73, ZCCM/7.1.4I, 'IRB MJIC, MUZ mtgs, mins, etc, 1970-74'.
80 Memorandum of Agreement for pay, leave and conditions 1/11/73-31/10/76, 28/2/74, ZCCM/3.1.4F, 'Mineworkers Union of Zambia, 1974'.
81 Draft Memorandum of Agreement, 2/11/73, ZCCM/3.1.4F, 'MUZ Negotiations, 1973-79'.
82 2nd PC MJIC mtg, 18/7/75, ZCCM/16.2.9B, 'MJIC-MUZ-ACME mtgs, 1974-75'.
83 3rd PC MJIC mtg, 29/7/75, Ibid.
84 Ibid.
85 David Mwila, address to SC, 14/6/76, in MUZ HO, 'Supreme Council minutes, 1972-76'.
86 Mindeco Year Book, 1973, ZCCM.
87 CISB telexes, 3/3/77 and 12/11/76, ZCCM/4.5.2I, 'Work stoppages, strikes, Jan 1972 – Sept 1978'.
88 Albert Chali interview, 4/2/2003.
89 Japhet Sekwila interview, 25/2/2003.
90 *ToZ*, 6/2/75.

91 See for example 4th PC MJIC mtg, 5/9/75, ZCCM/16.2.9B, 'MJIC-MUZ-ACME mtgs, 1974-75'.
92 Sam Kangwa interview, 5/10/2002.
93 *SToZ*, 13/7/75.
94 *ToZ*, 16/7/75.
95 Informal ACME-MUZ mtg, 17/6/75, ZCCM/16.2.9B, 'MJIC-MUZ-ACME mtgs, 1974-75'.
96 *ToZ*, 6/8/75.
97 CISB telex, 5/8/75, ZCCM/4.5.21, 'Work stoppages, strikes, Jan 1972 – Sept 1978'.
98 *ToZ*, 18/4/77.
99 *Ibid.*, 6/12/79.
100 Examples in MUZ Nkana/11B, 'MUZ Public activities and regional meetings'.
101 John Kapansa interview, 31/1/2003.
102 Enock Chansa interview, 11/2/2003.
103 George Mukoma interview, 11/2/2003.
104 Enock Chansa interview, 11/2/2003.
105 Asst Personnel Superintendent, MUZ Officials, and Women Representatives of Demonstrators, 15/3/78, in MUZ Luanshya, 'Meetings with Management, 1975-78'.
106 Ibid.
107 Rodah Matoba interview, 27/1/2003.
108 *SToZ*, 8/4/79.
109 *Ibid.*, 22/4/79.
110 *ToZ*, 24/10/78.
111 *Ibid.*, 2/10/79; see also Hamalengwa, *Class Struggles in Zambia*, p. 92.
112 *Ibid.*, 2/3/80.
113 *Ibid.*, 3/10/80, 30/9/80.
114 *Ibid.*, 2/10/80. 'Poland' is a reference to the Solidarity union movement, which was at that time in conflict (both industrial and political) with the Communist government.
115 DG Ndola Alexander Kamalondo, in *Ibid.*, 13/10/80.
116 *Ibid.*, 28/10/80.
117 *Ibid.*, 4/11/80.
118 *Ibid.*, 31/12/81.
119 CISB to Companies, Lusaka Directors of Personnel, 21/9/78, ZCCM/13.4.10E, 'MJIC, MUZ, ACME, 1975-80'.
120 SC, 28/6/78, in MUZ HO, 'Supreme Council & CEC minutes, 1975-81'.
121 MUZ Kabwe, n.d., received at Roan, 20/9/78, in MUZ Luanshya 6A, 'Luanshya JEC and caucus meetings, 1979-81'.
122 Sec Report to 1978 Conference – dated 19/9/78, in Ibid.
123 MUZ Biannual Conference, NC Opening Remarks, Kasama, 27-30/9/78, in MUZ Nkana, miscellaneous notes.
124 SC, 30/12/78, in MUZ HO, 'Supreme Council & CEC minutes, 1975-81'.
125 Record of Meeting Held at State House, 25/10/78 'SECRET', in MUZ HO, 'UNIP, Politics, 1977-89'.
126 Ibid.

127 CISB document, 'Dispute between the ACME and the MUZ', 13/2/79, ZCCM/1.3.1H, 'ACME-MUZ Dispute, 1979-88'.
128 Acting Superintendent Personnel Research to Manager IR, 6/4/79, 'Discussion Paper, MUZ Negotiations 1979', ZCCM/4.5.1C, 'MUZ Wage Negotiations, 1979'.
129 1st PC MJIC mtg, 21/3/80, ZCCM/13.4.10E, 'MJIC, MUZ, ACME, 1975-80'.
130 *ToZ*, 25/8/80.
131 CIRS, 'Zambian General Payroll Salaries: Policy Considerations', 1/7/81, ZCCM/4.5.1D, 'MUZ Negotiations, 1981'.
132 Charity M'tonga interview, 30/1/2003.
133 *ToZ*, 4/9/80.
134 Maxwell Kalesha-Chisoko interview, 6-7/2/2003.
135 1981 comparison of MIBPS with other Organisations, ZCCM/4.5.1D, 'MUZ Negotiations, 1981'.
136 Mijere, 'The Mineworkers' Resistance to Governmental Decentralisation in Zambia', pp. 3-4; see also Nordlund, *Organising the Political Agora*, pp. 71-77.
137 Burawoy, 'The Hidden Abode of Under-Development', pp. 123-124.
138 A.F. Greenwood, and L. Rowland, untitled paper on the integration of mine townships, 30/8/68, in MUZ HO IB/B/71, 'Other Government Ministries and Government Bodies, 1968-1981'.
139 Ibid.
140 Memo, 13/2/74, 'transport to and from work', ZCCM/16.2.9B, 'Group Industrial Relations Managers Job Evaluation Committee meetings, 1973-74'.
141 See Epstein, *Politics in an Urban African Community*; Powdermaker, *Copper Town*, both passim.
142 Charity M'tonga interview, 30/1/2003.
143 Mijere, 'The Mineworkers' Resistance to Governmental Decentralisation in Zambia', p. 188.
144 Ibid., p. 296.
145 MUZ SC, 22/7/75, in MUZ HO, 'Supreme Council minutes, 1972-76'.
146 Submission to Official Working Party on Mine Township Integration, 31/12/75, in MUZ HO IB/B/71, 'Other Government Ministries and Government Bodies, 1968-1981'.
147 J. Manda, Mufulira, 5/2/76, in Ibid.
148 Final submission to Official Working Party on Mine Township Integration, n.d., in Ibid.
149 *ToZ*, 19/1/76.
150 *Ibid.*, 24/1/76.
151 *Ibid.*, 22/1/76.
152 *Ibid.*, 3/2/76.
153 *Ibid.*, 4/2/76.
154 Manager Administration and MUZ Officials, 21/11/78, in MUZ Luanshya, 'Meetings with Management, 1978-80'.
155 *ToZ*, 21/8/80.
156 Mijere, 'The Mineworkers' Resistance to Governmental Decentralisation in Zambia', p. 303.
157 Ibid, pp. 307-8.
158 David Longwani interview, 11/2/2003.

159 Oswell Munyenyembe interview, 26/2/2003.
160 Sefelino Mumba interview, 15/2/2003.
161 Walamba interview.
162 Unnamed, quoted in Mijere, 'The Mineworkers' Resistance to Governmental Decentralisation in Zambia', p. 196.
163 *ToZ*, 22/8/81.
164 *Ibid.*, 18/10/80.
165 Mwila address to 8th Biannual Conference, Livingstone, 16-18/10/80, in MUZ Luanshya/5A, 'Biannual Conference, 1970-84'.
166 8th MUZ Conference Resolutions, in Ibid.
167 John Kapansa interview, 31/2/2003.
168 *ToZ*, 3/11/80.
169 MUZ NC Mwila to President Kaunda, 16/11/80, in MUZ HO, 'UNIP, Politics, 1977-89'.
170 SC mins 23 & 28/11/80, in MUZ HO, 'Supreme Council & CEC minutes, 1975-81'.
171 *ToZ*, 29/11/80.
172 SC mins 19/12/80, in MUZ Luanshya, 'MUZ Branch mtgs with Management, SC etc, 1979-81'.
173 *ToZ*, 10/1/81.
174 *Ibid.*, 12/1/81.
175 *Ibid.*, 17/1/81.
176 *Ibid.*, 21/1/81.
177 *Ibid.*, 24/1/81.
178 *Ibid.*, 22/1/81.
179 SC mins 8/5/81, in MUZ Luanshya, 'MUZ Branch mtgs with Management, SC etc, 1979-81'; *ToZ*, 24/1/81.
180 *ToZ*, 26/1/81.
181 George Mulenga interview, 6/3/2003.
182 *ToZ*, 26/1/81.
183 Nordlund, *Organising the Political Agora*, p. 76.
184 *ToZ*, 28/1/81.
185 *Ibid.*, 7/2/81.
186 *Ibid.*, 30/1/81.
187 *Ibid.*, 19/2/81.
188 *Ibid.*, 21/2/81.
189 *Ibid.*, 28/4/81.
190 *Ibid.*, 4/5/81.
191 Senior Personnel Officer 18 Shaft, SSs and WCs, 30/5/79, in MUZ Luanshya, 'Meetings with Management, 1978-80'.
192 Japhet Sekwila interview, 25/2/2003.
193 See for example *ToZ*, 8/7/81.
194 *SToZ*, 31/5/81.
195 Walamba interview.
196 Ibid.
197 *ToZ*, 1/7/81.
198 *Ibid.*, 1/6/81.
199 John Chalwe interview, 12/2/2003.

200 *ToZ*, 8/7/81.
201 Peter Chileshe interview, 12/12/2003.
202 George Mukoma interview, 11/2/2003.
203 *ToZ*, 9/7/81.
204 *ToZ*, 10/7/81; SC mins, 14/7/81, in MUZ Luanshya, 'MUZ Branch mtgs with Management, SC etc, 1979-81'.
205 *ToZ*, 16/7/77.
206 Personnel Superintendent, MUZ Officials and Works Councillors, 3/6/81, in MUZ Luanshya/6C, 'Mtgs with local management, 1980-83'.
207 'Forthcoming MUZ Negotiations, visit to Chibuluma', 28/3/79, ZCCM/4.5.1.C, 'MUZ Wage Negotiations, 1979'.
208 Manjolo Banda interview, 14/2/2003; supported by George Mukoma, 11/2/2003; Stephen Musonda, 29/1/2003.
209 Percy Chanda interview, 27/3/2003.
210 Petition, 'Workers' to MUZ, CISB, and the Ministry of Labour and Social Services, 12/6/81, in MUZ Luanshya untitled file, grievance cases, correspondence, 1970-81.
211 Interviews: Mary Palesi, 7/3/2003; Rodrick Chipili, 25/2/2003; Percy Chanda, 27/2/2003.
212 George Mulenga interview, 6/3/2003.
213 *ToZ*, 1/7/81.
214 *Ibid.*, 16/7/81.
215 *Ibid.*, 18/7/81.
216 *Ibid.*
217 *SToZ*, 19/7/81.
218 For example David Longwani interview, 11/2/2003.
219 *ToZ*, 4/12/81.
220 Nordlund, *Organising the Political Agora*, p. 78.
221 Walamba interview.
222 *ToZ*, 28/7/81.
223 *Ibid.*, 1/8/81; Hamalengwa states that the ZCTU discussed calling for a general strike in protest, but instead issued an appeal to President Kaunda for the leaders to be released: *Class Struggles in Zambia*, p. 102.
224 *ToZ*, 18/9/81.
225 *Ibid.*, 7/12/81.
226 Nordlund, *Organising the Political Agora*, p.80.

Chapter Five

1 K. Woldring, 'Survey of Recent Inquiries and their Results', in *Beyond Political Independence: Zambia's Development Predicament in the 1980s*, ed. K. Woldring, Mouton Publishers (Berlin, 1984), p. 205.
2 Akwetey, *Trade Unions and Democratisation*; Nordlund, *Organising the Political Agora*; Rakner, *Trades Unions in Processes of Democratisation.*
3 Nordlund, *Organising the Political Agora*, p. 297.
4 See for example P. Gibbon, 'Structural Adjustment and Pressures Towards Multi-Partyism in sub-Saharan Africa' in Y. Bangura, P. Gibbon and A. Ofstad (eds.), *Authoritarianism, Democracy and Adjustment: the Politics of Economic*

Reform in Africa, Nordiska Africainstitutet (Uppsala, 1992); cited in Rakner, *Trades Unions in Processes of Democratisation*, p. 114.

5 R.H. Bates and P. Collier, 'The Politics and Economics of Policy Reform in Zambia', in *Political and Economic Reform: Evidence from Eight Countries*, ed. R.H. Bates and A.O. Krueger, Blackwell (Oxford and Cambridge MA, 1993); L. Rakner, *Political and Economic Liberalisation in Zambia, 1991-2001*, Nordiska Africainstitutet (Uppsala, 2003).

6 Rakner, *Trades Unions in Processes of Democratisation*, passim.

7 *ToZ*, 23/10/82.

8 See Zimba's demand that wage rises match a 100% rise in prices: *ToZ*, 14/1/82.

9 ZCTU GC meeting minutes, 30/12/83, in MUZ HO, 'ZCTU-related papers, 1977-92'.

10 *ToZ.*, 1/6/83.

11 *Ibid.*, 2/8/83.

12 *SToZ*, 7/10/84.

13 *Ibid.*, 1/11/84.

14 *Ibid.*, 3/8/84.

15 *Ibid.*

16 Patu Simoko, in *SToZ*, 5/8/84.

17 *ToZ*, 5/2/84.

18 *Ibid.*, 21/8/84.

19 *ToZ*, 3/1/83.

20 *Ibid.*, 14/2/84.

21 CEC, 20/8/82, in MUZ Nkana, 'MUZ SC and CEC meetings, 1970-89'.

22 Ibid.

23 *ToZ*, 3/5/84.

24 *Ibid.*, 27/4/85.

25 *ToZ*, 21/1/85; Nordlund, *Organising the Political Agora*, p. 87.

26 *ToZ*, 5/3/85, 12/3/85.

27 *Ibid.*, 26/3/85.

28 E. Beele, 'The Bolting Workers: Some Scenarios of Statutory Instrument Number 6 of 1985' *Zambia Journal of Business*, 4, 1 & 2 (Lusaka, 1985), cited in Hamalengwa, *Class Struggles in Zambia*, pp. 27-29.

29 *ToZ*, 5/11/84, 4/12/84. See also Nordlund, *Organising the Political Agora*, pp. 62-63.

30 *ToZ*, 19/7/85.

31 2nd Ordinary mtg of Tripartite Labour Consultative Council, 18-20/9/88, ZCCM/8.6.1D, 'Labour Consultative Committee, Labour Department reports, 1989-2000'.

32 N.L. Zimba, address to SC, 30/9/88, in MUZ HO, 'Supreme Council minutes, 1983-88'.

33 ZCTU, 'ZCTU's Views on the Impending Come Back of the IMF and "Structural Adjustment with a Human Face", n.d. (1989), in MUZ HO, Miscellaneous papers.

34 *ToZ*, 27/2/82.

35 For example interviews with Oswell Munyenyembe, 26/2/2003; Peter Chileshe, 12/2/2003; Nordlund, *Organising the Political Agora*, p. 62, p. 82.

36 *SToZ*, 28/2/82.
37 *ToZ*, 6/3/82.
38 *SToZ*, 16/5/82.
39 MUZ SC, 21/8/82, in MUZ HO, 'Supreme Council minutes, 1981-87'.
40 *ToZ*, 22/5/82.
41 2nd MJIC mtg, 16/5/86, ZCCM/16.2.9C, 'GIRM Committee and related correspondence, 1986'.
42 *ToZ*, 1/10/82.
43 *Ibid.*, 29/9/82.
44 Harrison Nkonde interview, 1/3/2003; Also John Chalwe interview, 12/2/2003.
45 George Lombe interview, 27/2/2003; Also Patrick Kanyanta interview, 7/11/2002.
46 *ToZ*, 29/3/84.
47 Complaint by R.E. Mapulanga, n.d. c. June 1982, in MUZ Luanshya, 'General Correspondence, 1980-85'.
48 Memo, All Check Point Clerks to Personnel Dept, 3/10/83, in Ibid.
49 Memo, CSSs and SSs Underground Mining Dept to Mgt ZCCM Nchanga Div, 30/3/90, in MUZ Nchanga, miscellaneous documents.
50 MJIC agreement, signed 19/8/83, for Aug 1983 to July 1985, ZCCM/1.9.4D, 'ACME, MJIC meetings, 1983-87'.
51 MUZ Press Conference statement, 14/12/83, 'ZCCM Should Not Take MUZ for a Ride', in MUZ Nchanga/5B, 'Supreme Council, 1979-85'.
52 *ToZ*, 15/2/85.
53 3rd MJIC meeting, 23/4/85, ZCCM/1.9.4D, 'ACME, MJIC meetings, 1983-87'.
54 Ibid.
55 5th MJIC mtg, 17/5/85, ZCCM/1.9.4D, 'ACME, MJIC meetings, 1983-87'.
56 For example, Central Services Department strike, reported in *SToZ*, 15/1/84.
57 Charity M'tonga interview, 30/1/2003.
58 Mimbula-Fitula sub-Branch public mtg, Kaunda Square, 24/2/85, in MUZ Nchanga/11B, 'Public meetings, 1972-86'.
59 Cameron Pwele interview, 17/3/2003.
60 Basilio Kunda interview, 7/2/2003.
61 George Mukoma interview, 11/2/2003.
62 For example, letter from Kalulushi miner, *ToZ*, 24/3/84.
63 SC, 22/6/84, in MUZ HO, 'MUZ Supreme Council minutes, 1981-87'; *ToZ*, 7/5/84.
64 Court Ruling, Lusaka, 23/4/84, ZCCM/1.9.4D, 'ACME, MJIC meetings, 1983-87'.
65 Executive Committee meeting, 7/9/82, in MUZ Nchanga, 'Executive Committee meetings, 1982-87'.
66 CEC meeting, 6/2/84, in MUZ Nchanga, 'Central Executive Committee meetings, 1978-85'.
67 Petition, 11/10/84, in MUZ HO, 'Supreme Council Minutes, 1983-88'; *ToZ*, 12/10/84; *SToZ*, 11/11/84.
68 Robert Kapenda interview, 22/1/2003.
69 Reported in *ToZ*, 19/10/84.

70 *ToZ*, 14/11/84.
71 David Ndalama interview, 13/2/2003.
72 Emergency MUZ SC, 16/11/84, in MUZ HO, 'Supreme Council minutes, 1981-1987'.
73 *ToZ*, 11/1/86.
74 B.J. Ngulube interview, 5/11/2002.
75 Boniface Kumwenda interview, 2/11/2002; Also interviews with Oswell Munyenyembe, 26/2/2003; Cameron Pwele, 17/3/2003; David Ndalama, 13/2/2003.
76 *ToZ*, 22/3/82, 8/5/82.
77 Mary Palesi interview, 7/3/2003.
78 David Longwani interview, 13/2/2003.
79 George Mukoma interview, 27/2/2003.
80 See for example, Boniface Kumwenda interview, 2/11/2002.
81 Mimbula-Fitula sub-Branch public mtg, 24/2/85, in MUZ Nchanga/11B, 'Public meetings, 1972-86'.
82 SC, 13/5/85, in MUZ HO, 'Supreme Council minutes, 1983-88'; MUZ Committee to review Mukuba Pension Scheme, mtg 18/5/85, in MUZ HO, 'Supreme Council minutes, 1981-87'.
83 *ToZ*, 26/5/85; GS MUZ to BS Nchanga, 25/3/85, in MUZ Nchanga/6A, 'Branch Executive correspondence, 1980-87'.
84 *ToZ*, 2/6/85.
85 *Ibid.*
86 Mary Palesi interview, 7/3/2003.
87 Harrison Nkonde interview, 1/3/2003.
88 ACME – MUZ special mtg, 1/6/85, ZCCM/1.9.4D, 'ACME, MJIC meetings, 1983-87'.
89 SC, 13/6/85, in MUZ HO, 'Supreme Council minutes, 1981-1987'.
90 Kramer Nyondo interview, 13/9/2002.
91 *ToZ*, 6-8/6/85.
92 Harrison Nkonde interview, 1/3/2003.
93 *ToZ*, 13/7/85.
94 Letter, J. Hoatson, GM Nchanga Division ZCCM to CIRS CISB, 22/2/85, ZCCM/1.5.3G, 'Industrial Relations Act 1971, 1970-1995'.
95 NC Opening Address to SC 27/12/85, in MUZ HO, 'Supreme Council minutes, 1983-88'; Also *ToZ*, 5/8/85.
96 *ToZ*, 6/9/85, 27/8/86.
97 ZCTU Secretary General's Opening Remarks to High-Level Trade Union Leadership Seminar, Kabwe, 25/11/85, in MUZ HO Research Department files.
98 Chiluba, speech to 1st ZCTU General Council, early 1986.
99 *ToZ.*, 15/5/86.
100 *Ibid.*, 11/10/86, 29/10/86.
101 *Ibid.*, 4/12/86.
102 *Ibid.*, 10/12/86.
103 *Ibid.*, 24/12/86.

104 *ToZ*, 20/11/85; amended agreement, signed 29/11/86, to existing agreement 1/5/85-30/4/87, ZCCM/1.4.9D, 'ACME, MJIC meetings, 1983-87'.
105 N. Simutanyi, 'Organised Labour, Economic Crisis and Structural Adjustment in Africa: The Case of Zambia', in *Democracy in Zambia: Challenges for the Third Republic* ed. O. Sichone and B.C. Chikulo, SAPES Books (Harare, 1996), p.164.
106 Roy Yumbe interview, 7/11/2002.
107 Interview, B.J. Ngulube, 5/11/2002. Also interviews with Boniface Kumwenda, 2/11/2002; Mwansa Chipotyo, 5/2/2003; Kramer Nyondo, 13/9/2002.
108 West, 'The Politics of Hope', p. 383; Nordlund, *Organising the Political Agora*, p.88.
109 Sefelino Mumba interview, 15/2/2003.
110 *ToZ*, 10/12/86.
111 *Ibid.*, 15/12/86.
112 *Ibid.*, 18/12/86.
113 *SToZ*, 1/2/87.
114 Agnes N'gandwe interview, 26/2/2003.
115 *ToZ*, 5/10/89.
116 *Ibid.*, 10/10/89.
117 CISB, 'Summary of Protests by Women at Nkana Division', 7/10/89, ZCCM/1.7.1E, 'GIRMs papers, 1989-90'; see also *ToZ*, 11/10/89.
118 Kitwe Senior DG, Senior MET and Nkana Division officials, Nkana and Mindolo MUZ mtg 9/10/89, 'to discuss grievances and allegations of malpractices in the Mine Hospitals as raised by demonstrating housewives', in MUZ Nkana, 'Branch officials meetings with management, 1976-1990'.
119 Emergency BEC, 7/10/89, in MUZ Nkana, 'Complaint Files, 1985-87'.
120 *ToZ*, 17/10/89.
121 MJIC mtg, 20/4/90, ZCCM/16.3.2C, 'MUZ-ACME, 1990-92'.
122 MJIC mtg mins, 31/10/89, ZCCM/1.7.1E, 'GIRMs Papers, 1989-90'.
123 Interviews: B.J. Ngulube, 5/11/2002; Boniface Kumwenda, 2/11/2002.
124 Enock Chansa interview, 11/2/2003; confirmed by Manjolo Banda interview, 14/2/2003.
125 Evelyn Musonda interview, 27/12/2002.
126 Fidelis Mwamba interview, 15/9/2002.
127 Harrison Nkonde interview, 1/3/2003.
128 Japhet Sekwila interview, 25/2/2003.
129 George Lombe interview, 27/2/2003.
130 George Mukoma interview, 11/2/2003.
131 Joseph Mulenga interview, 8/11/2002.
132 Ibid.
133 Boniface Kumwenda interview, 2/11/2002; Government Gazette Notice No 819, 1988, quoted in ZCCM Corporate Head Office MPU Act, memorandum, 23/3/90, ZCCM/6.6.2H, 'Luanshya Mine Police Unit reports, May 1990 – June 1991'.
134 Ibid.
135 See cases recorded in ZCCM/6.6.2H, 'Luanshya Mine Police Unit reports, May 1990 – June 1991'.

136 SC, 15/7/88, in MUZ HO, 'Supreme Council minutes, 1983-88'; *ToZ*, 21/4/88.
137 See for example, NC opening remarks to SC, 15/7/88, in MUZ HO, 'Supreme Council minutes, 1983-88'.
138 See for example Akwetey, *Trade Unions and Democratisation*, pp. 56-7.
139 Frederick Hapunda, Minister of Labour 1982-86, quoted in Nordlund, *Organising the Political Agora*, p. 112.
140 Walamba interview.
141 SC, 30/9/88, in MUZ HO, 'Supreme Council minutes, 1983-88'.
142 Agnes N'gandwe interview, 26/2/2003.
143 Boniface Kumwenda interview, 2/11/2002.
144 Edward Kabundi interview, 18/2/2003.
145 Borries Ntankula interview, 5/11/2002.
146 *ToZ*, 19/7/89.
147 *ToZ*, 20/7/89.
148 *Ibid.* ZCBC was a state-owned store. Also Annie Chamululu interview, 3/2/2003.
149 *ToZ*, 9/8/89.
150 *Ibid.*, 8/8/89.
151 *Ibid.*, 30/12/89.
152 *ToZ*, 17/8/89.
153 Meeting of the Labour Officers, 14/4/88, NAZ, MoL unnumbered file, 'Tripartite Consultative Council meetings, 1988'.
154 Meeting in Office of SoS for Defence and Security, 6/4/88, Ibid.
155 Ibid.
156 Hamalengwa, *Class Struggles in Zambia*, p. 106.
157 *ToZ*, 16/6/90.
158 *Ibid.*, 1/8/90.
159 *Ibid.*, 6/10/90.
160 George Lombe interview, 27/2/2003.
161 Albert Chali interview, 4/2/2003; *ToZ*, 5/9/90; Robert Kapenda interview, 22/2/2003.
162 Timothy Walamba interview, 6/11/2002.
163 Evelyn Musonda interview, 27/12/2003. 'Simakuni' was likened to 'inkuni', Bemba for 'firewood', symbolising that he was a 'firebrand' union leader. The use of a CiBemba term for Simakuni, a Tonga, is indicative of the cross-ethnic discourse prevalent in MUZ.
164 Ibid.
165 George Mulenga interview, 6/3/2003.
166 Mary Palesi interview, 7/3/2003.
167 George Mukoma interview, 11/2/2003.
168 MUZ 1990 Biannual [sic] Conference, Report to Nkana Main Branch Executive, in MUZ Nkana, Miscellaneous documents.
169 *ToZ*, 19/3/90; MUZ 1990 Biannual Conference, in MUZ Mufulira/5/A/2/90 – 'Biennial Conference, Correspondence file, 1990-1998'.
170 Nordlund, *Organising the Political Agora*, p. 99.
171 *ToZ*, 22/3/90.

172 CISB memo, Mgr IR N. Simukonda to Snr Mgr IR, 11/4/90, ZCCM/16.3.2C, 'MUZ-ACME, 1990-92'.
173 See for example, BEC mtg, 30/1/90, 7/2/90, in MUZ Nkana, 'Main Executive Committee meetings, 1990-91'.
174 Steven Mukuka interview, 18/3/2003.
175 *ToZ*, 24/3/90.
176 MUZ SC 16/10/90, J. Simakuni, opening remarks, in MUZ HO, miscellaneous papers.
177 *ToZ*, 24/8/90.
178 *SToZ*, 4/11/90.
179 *ToZ*, 6/11/90.
180 *Ibid.*, 19/12/90.
181 Albert Chali interview, 4/2/2003.
182 'The Hour' was the MMD's slogan; its use in this context demonstrates the politicised nature of this dispute.
183 Sefelino Mumba interview.
184 Ibid.
185 Nchanga Main MUZ, n.d., list of those dismissed in Nov 1990 strikes, in MUZ Nchanga/17A, 'CISB correspondence, 1987-91'; Sefelino Mumba interview.
186 Sefelino Mumba interview.
187 Special MJIC, 26/6/90, in MUZ HO, 'MJIC papers, 1985-91'.
188 8th MJIC, 28/11/90, in Ibid.
189 Ibid.
190 *SToZ*, 2/12/90.
191 Interviews: Agnes N'gandwe, 26/2/2003, Peter Lusaka, 18/2/2003; Nkana BEC, 15/3/91, in MUZ Nkana, 'Main Executive Committee meetings, 1990-91'.
192 George Mukoma interview, 11/2/2003.
193 MUZ Press Release, GS K. Sheng'amo, n.d., following CEC meeting, 8/1/91, in MUZ Nchanga, miscellaneous documents.
194 e.g. in Ndola: *SToZ*, 13/1/91.
195 *ToZ*, 4/5/91.
196 *SToZ*, 5/5/91.
197 Boniface Kumwenda interview, 5/11/2002.
198 Mary Palesi interview, 7/3/2003.
199 Kramer Nyondo interview, 13/9/2002.
200 *ToZ*, 8/5/91.
201 Harrison Nkonde interview, 1/3/2003.
202 David Longwani interview, 11/2/2003.
203 David Ndalama interview, 13/2/2003.
204 *Sunday Express* (Zambia), 21/7/91.
205 *ToZ*, 27/12/90.
206 *Ibid.*, 20/6/91.
207 Akwetey, *Trade Unions and Democratisation*, p. 102.
208 *SToZ*, 30/12/90.
209 CEC minutes, 8/1/91, in MUZ HO, 'CEC minutes and related papers, 1986-1997'

210 *ToZ*, 15/2/83.
211 *Ibid.*, 6/2/91.
212 Percy Chanda interview, 27/3/2003.
213 Ibid.
214 *ToZ*, 6/6/90.
215 *Ibid.*, 27/10/90.
216 *Ibid.*, 16/2/91.
217 *Ibid.*, 18/5/91.
218 *Ibid.*, 1/5/91.
219 Speaking at UNIP rally in Kitwe, reported in *ToZ*, 23/8/90.
220 *ToZ*, 8/8/91.
221 Akwetey, *Trade Unions and Democratisation*, p. 65.
222 Nordlund, *Organising the Political Agora*, pp. 104-5.
223 *ToZ*, 31/10/91.
224 *SToZ*, 24/11/91.
225 John Bwalya interview, 5/2/2003; supported by Mwansa Chipotyo interview, 5/2/2003.
226 Abel Chisanga interview, 5/11/2002.
227 Harrison Nkonde interview, 1/3/2003.
228 Maxwell Kalesha-Chisoko interview, 6-7/2/2003 .
229 Evelyn Musonda interview, 27/12/2002.
230 Albert Chali interview, 4/2/2003.
231 Sam Kangwa interview, 5/10/2002.
232 *ToZ*, 24/1/91, 31/1/91.
233 *SToZ*, 28/4/91.
234 *Ibid.*, 27/10/91.
235 Albert Chali interview, 4/2/2003; also Peter Lusaka interview, 15/2/2003.
236 Boniface Kumwenda interview, 2/11/2002; also Peter Chileshe interview, 12/2/2003.
237 Agnes N'gandwe interview, 26/2/2003.
238 *ToZ*, 30/10/91.
239 Abel Chisanga interview, 5/11/2002.
240 Interviews: Oswell Munyenyembe, 26/2/2003; Raphael Mwanza, 1/3/2003; Kramer Nyondo, 13/9/2002; Patrick Kanyanta, 10/11/2002.
241 M. Bratton, 'Economic Crisis and Political Realignment, Zambia', in J. Widner (ed.) *Economic Change and Political Liberalization in sub-Saharan Africa*, John Hopkins Press (Baltimore, 1994).
242 George Lombe interview, 27/2/2003.
243 MJIC, 4/10/91, in MUZ HO, 'MJIC papers, 1985-91'.
244 MJIC, 11/10/91, in Ibid.
245 MJIC, 5/11/91, in Ibid.
246 *ToZ*, 26/11/91.
247 Opening remarks to MUZ SC, 25/11/91, F Kunda, in MUZ HO, miscellaneous papers.

Chapter Six

1 Rakner, *Political and Economic Liberalisation in Zambia*, pp. 78- 9.

2 M. Tshoaedi, 'Trade Unions in Zambia', *South Africa Labour Bulletin*, 24, 3 (2000).
3 Zambia – European Community, *Country Strategy Paper and National Indicative Programme for 2001 – 2007* (2001).
4 G. Koyi, 'Impact of Structural Adjustment on Health in Zambia', in *The Impact of Structural Adjustment Programmes on the Zambian Economy: A Trade Union Perspective*, Zambia Congress of Trade Unions/International Confederation of Free Trade Unions-African Regional Organisation Research Project (Kitwe, 2001), p. 76.
5 *Ibid.*, p. 44.
6 Inter-African Network for Human Rights & Development (Afronet)/Citizens for a Better Environment (CBE)/Rights and Accountability in Development (RAID), *Zambia: Deregulation and the denial of human rights. Submission to the OECD Committee on Economic, Social and Cultural Rights*, (Lusaka, Kitwe, Oxford, 2000), p. 2.
7 Rakner, *Political and Economic Liberalisation in Zambia*, p. 85.
8 Zambia – European Community, *Country Strategy Paper*, p. 4.
9 Afronet/CBE/RAID, *Zambia*, p. 61.
10 Transparency International Zambia, *Transparency and The Rule of Law in The Privatization of The Zambia Consolidated Copper Mines (ZCCM) Assets* (Lusaka, 2002).
11 Afronet/CBE/RAID, *Zambia*, pp. 59-63.
12 Rakner, *Political and Economic Liberalisation in Zambia*, p. 103.
13 *Ibid.*, pp. 132-133.
14 Personal communications in Lusaka and Kitwe, August 2001.
15 For example, the European Union Election Observation Mission, see www.eueu-Zambia.org for reports.
16 *The Post*, 6/8/2003.
17 For example Abel Chisanga interview, 5/11/2002.
18 Raphael Kasonde interview, 23/2/2003. Similar sentiments are expressed by Mwansa Chipotyo (interview, 5/2/2003), Japhet Sekwila (25/2/2003).
19 Borries Ntankula interview, 5/11/2002.
20 President Chiluba to General Secretaries of trade unions, n.d. (December 1991), in MUZ HO/15/13/81, 'Other Government Ministries and Government Bodies, 1981-2001'.
21 Rakner, *Economic and Political Liberalisation in Zambia*, p. 95.
22 *Ibid.*
23 Afronet/CBE/RAID, *Zambia*, pp. 97-8.
24 Tshoaedi, 'Trade Unions in Zambia', pp. 88-89.
25 This point is made by Austin Muneku, ZCTU Director of Research, various discussions 2001 – 2002.
26 ZCTU/ICFTU-AFRO, *Impact of Structural Adjustment Programmes.*
27 Leonard Hikaumba interview, 10/3/2003.
28 *Sunday Post*, 25/1/2003.
29 Telephone discussion with Joyce Nonde, ZUFIAW General Secretary, 2/4/2003.
30 EC-Zambia, *Country Strategy Paper*, p. 8; *The Post*, 19/12/2002.
31 *ZNBC TV*, 10/4/2005.

32 Advertisement in *The Post*, 12/4/2005.
33 *The Post*, 11/8/2005.
34 *The Post*, 18/7/2003.
35 *The Post*, 19/2/2004.
36 This is a conclusion of the Afronet/RAID/CBE report. I am indebted to RAID's Patricia Feeney for additional information on this subject.
37 Discussions with senior MUZ officials, Kitwe, Oct 2002 – Mar 2003; *ToZ*, 13/9/98, cited in Afronet/CBE/RAID, *Zambia*, p. 81.
38 Rakner, *Political and Economic Liberalisation*, p. 77; J.R. Craig, 'State Enterprise and Privatisation in Zambia, 1968 – 1998', PhD Thesis, University of Leeds (1999), pp. 171-174.
39 F. Kaunda, *Selling the Family Silver: the Zambian Copper Mines Story*, Printpak Books (KwaZulu Natal, 2002), p. 151.
40 Rakner, *Political and Economic Liberalisation*, p. 78.
41 Afronet/CBE/RAID, *Zambia*, p. 173.
42 Discussions with former senior Binani executive, unattributable, Ndola, 2003.
43 Ibid.
44 Interviews with Henry Mukuka, 16/12/2002; Cameron Pwele, 17/3/2003.
45 Kelvin Kachingwe, article in *ToZ*, 23/1/2003.
46 Discussions with George Mulenga, MUZ Luanshya Organising Secretary, December 2002.
47 Cameron Pwele interview, 17/3/2003.
48 Ibid.
49 For example Agnes N'gandwe interview, 26/2/2003.
50 Interviews: Charity M'tonga, 30/1/2003, Peter Chisanga, 14/2/2003; discussions with Charles Muchimba, MUZ Director of Research, 2002 - 2003.
51 A.K. Nguluwe interview, 6/11/2002; also Albert Chali interview, 4/2/2003.
52 Boniface Kumwenda interview, 2/11/2002.
53 George Mukoma interview; also Annie Chamululu interview, 3/2/2003.
54 Afronet/CBE/RAID, *Zambia*, p. 74.
55 Mwansa Chipotyo interview, 5/2/2003.
56 Percy Chanda interview, 27/3/2003; also Borries Ntankula interview, 5/11/2002.
57 *ZNBC*, 3/8/2005
58 *The Post*, 26/4/2005; bbc.co.uk, 25/4/2005.
59 *The Post*, column by Sikota Wina, 26/4/2005.
60 *The Post*, 25/7/2005.
61 *Zambia Daily Mail*, 18/8/2005.
62 Charity M'tonga interview, 30/1/2003. Supported by Kramer Nyondo interview, 13/9/2002. Such views of village life amongst retiring mineworkers are provided in Ferguson, *Expectations of Modernity*, passim.
63 For example Albert Chali interview, 4/2/2003.
64 Evelyn Musonda interview, 27/12/2002. Similar views are expressed by Henry Mukuka, interview, 16/12/2002.
65 Boniface Kumwenda interview, 2/11/2002.
66 David Longwani interview, 11/2/2003; also interviews with Abel Chisanga, 5/11/2002; George Mukoma, 11/2/2003.
67 George Mulenga interview, 6/3/2003.

68 John Bwalya interview, 5/2/2003.
69 Patrick Kanyanta interview, 7/11/2002.
70 Raphael Mwanza interview, 1/3/2003.
71 Pascale Mwansa interview, 26/2/2003; also Manjolo Banda interview, 14/2/2003.
72 Oswell Munyenyembe interview, 26/2/2003.
73 Timothy Mfula interview, 7/11/2002; also Stephen Musonda interview, 29/1/2003.
74 A.K. Nguluwe interview, 6/11/2002.
75 Edward Kabundi interview, 18/2/2003. Other interviewees expressing similar views: Pacale Mwansa, 26/2/2003; Lukias Banda, 08/2/2003; Anton Kaluba, 16/12/2002; Percy Chanda, 27/3/2003.
76 Boniface Kumwenda interview, 2/12/2002.
77 Albert Chali interview, 4/2/2003.
78 David Ndalama interview, 13/2/2003. Fenwick Nyirongo (interview,10/2/2003), George Lombe (27/2/2003) expressed similar views.
79 Timothy Mfula interview, 7/11/2002.

Conclusion

1 Hamalengwa, *Class Struggles in Zambia*, pp. 55 – 56.
2 Burawoy, *The Colour of Class on the Copper Mines*, p. 80.
3 Cooper, *Decolonization and African Society*, p. 328.
4 E.P. Thompson, *The Making of the English Working-Class*, Penguin (London, 1963, 1968, 1980), Preface, p.10.
5 R. Michels, *Political Parties: a Sociological Study of the Oligarchical Tendencies of Modern Democracy*, Glencoe (1911); also R. Hyman, *Marxism and the Sociology of Trade Unionism*, Pluto Press (London, 1971, 1973). Both cited in Campbell, *The Scottish Miners, Volume Two:*), Introduction.
6 Campbell, *The Scottish Miners, 1874 – 1939*, passim.
7 Ferguson, *Expectations of Modernity*, passim.
8 M. Bratton, *The Local Politics of Rural Development: Peasant and Party-State in Zambia*, University Press of New England (Hanover NH and London, 1980), passim.
9 During the 2001 election, a special edition of the *Times of Zambia* was sold in Northern and Luapula Provinces whilst voting was still taking place. It claimed that the MMD had won in the Copperbelt. This aimed to influence voters in these provinces to follow their urban counterparts, on the basis that the latter were better informed about political issues. Some respondents argued that this influences voting behaviour in northern Zambia, particularly Albert Chali, interview, 4/2/2003.
10 Burawoy, *The Politics of Production*, passim.

APPENDIX 1: INTERVIEWEES

This appendix provides details on each of the 62 former local and national leaders of the Mineworkers' Union of Zambia (MUZ) interviewed for this study in 2002-3. This Appendix gives a personal context to the respondents, by providing personal, employment, and union career information for each interviewee. Unless otherwise stated, they remain in employment in the mines.

Lukias Banda
Male. *Year of birth*: 1954. *Place of birth*: Chipata, Eastern Province (hereafter EP).
Year of employment: 1976. *Place(s) of work*: Nchanga mine, Medical Department.
Jobs: Lasher; Machine Man; Sprayman; Storeman Clerk; Laundryman; Phyisotherapy Assistant; Clerk; Ambulance Controller; Ambulance Driver. Dismissed in 2001.
MUZ positions: 1981, Shop Steward. 1985-2001, Chief Shop Steward.

Manjolo Banda
Male. *Year of birth*: 1958. *Place of birth*: unknown area, EP.
Year of employment: 1979. *Place(s) of work*: Kalulushi and Konkola mines.
Jobs: Lasher; Driller Helper; Loader Driver; Blasting Licence Holder; General Timberman; Section Boss; Materials Distributor; Materials Contributor.
MUZ positions: 1986-91, 1993–97, Shop Steward.

John Bwalya
Male. *Year of birth*: 1944. *Place of birth*: Samfya, Luapula Province (LP).
Year of employment: 1971. *Place(s) of work*: Konkola mine hospital.
Jobs: Cleaner; Record Clerk. Retrenched 1993.
MUZ positions: 1976, Shop Steward. 1985-93, Chief Shop Steward.

Albert Chali
Male. *Year of birth:* 1951. *Place of birth:* Samfya, LP.
Year of employment: 1973. *Place(s) of work:* Nchanga mine.
Jobs: Helper; Boiler Maker; Leading Mechanic. Retrenched 1992.
MUZ positions: 1982, Branch Vice Chairman. 1986–90, Branch Chairman.

John Chalwe
Male. *Year of birth:* 1952. *Place of birth:* Mansa, LP.
Year of employment: 1971. *Place(s) of work:* Konkola mine and mine hospital.
Jobs: Lasher; Bell Man; Drilling Helper; Blasting Licence Holder; Machine Operator; Records Clerk; Senior Records Clerk, mine hospital.
MUZ positions: 1988, Shop Steward – to ?.

Annie Chamululu
Female. *Year of birth:* 1952. *Place of birth:* Chingola, Copperbelt (CBP).
Year of employment: 1976. *Place(s) of work*: Mufulira mine hospitals and clinics.
Jobs: Midwife; Nurse; Head Enrolled Nurse.
MUZ positions: 1988, Shop Steward. 1989, Branch Vice Treasurer. 1990, Treasurer. 2000, Branch Secretary.
1990–99, ZCTU Women's Committee, Vice Chairperson part of this time.

Percy Chanda
Male. *Year of birth:* 1955. *Place of birth:* unknown area, Northern Province (NP).
Year of employment: 1978. *Place(s) of work:* Kabwe mine, Mufulira mine.
Jobs: Artisan Boilermaker, Senior Artisan.
1980–81: Kabwe Mine Works Council Chairman.
1990–91: Pro MUZ National Chairman.
MUZ positions: 1994-98, Mufulira Main Branch Chairman.

Enock Chansa
Male. *Year of birth:* 1946. *Place of birth:* Kitwe, CBP.
Year of employment: 1975. *Place(s) of work:* Konkola.
Jobs: General worker, underground and surface. Retired 2000(?).
MUZ positions: 1969-75, Konkola Branch office worker. 1975, Shop Steward. 1978-88, Vice Treasurer. 1988-90, Treasurer. 1990-98, Chief Shop Steward. 1998-2000, Treasurer.

Peter Chileshe
Male. *Year of birth:* 1950. *Place of birth:* Kitwe, CBP.
Year of employment: 1971. *Place(s) of work:* Konkola mine.
Jobs: Lasher; Helper; Bellhand; Jackhammer Machine Helper; Blasting Licence Holder; Timberman; Box Timberman; Section Boss; Shift Boss. Retired 1997.
MUZ positions: 1976, Shop Steward. 1982-90, Chief Shop Steward.

Rodrick Chipili
Male. *Year of birth:* 1952. *Place of birth:* Chingola, CBP.
Year of employment: 1974. *Place(s) of work:* Mufulira mine.
Jobs: Workman; Whistleman; Loco Driver; Disc Clerk; Lift Clerk; Assessment Officer; Assistant Training Officer; Training Officer.
MUZ positions: 1989, Mufulira Main Branch Vice Treasurer. 1993, Treasurer. 1997-98, Branch Chairman.

Mwansa Chipotyo
Male. *Year of birth:* 1954. *Place of birth:* Kawambwa, LP.
Year of employment: 1974. *Place(s) of work:* Konkola mine.
Jobs: Apprentice; Artisan; Pump Chamber Supervisor. Medical discharge 2002.
MUZ positions: 1988, Shop Steward. 1990, Konkola Branch Vice Secretary. 1995, Branch Vice Chairman.

Abel Chisanga
Male. *Year of birth:* 1950. *Place of birth:* Mansa, LP.
Year of employment: 1970. *Place(s) of work:* Main Shaft, Nkana.
Jobs: Workman. *MUZ positions:* 1992, Shop Steward, to date.

Peter Chisanga
Male. *Year of birth:* 1951. *Place of birth:* Mporokoso, NP.
Year of employment: 1970. *Place(s) of work:* Konkola mine and concentrator.
Jobs: Lasher; Workman; Concentrator Operator.
MUZ positions: 1975, Shop Steward. 1979, Chief Shop Steward. 1985-86, Konkola Branch Vice Secretary. 1987-98, Shop Steward.

John Chisata
Male. *Year of birth:* 1935. *Place of birth:* Chinsali, NP.
Year of employment: 1953. *Place(s) of work:* Mufulira mine. *Jobs:* Unknown.
AMWU positions: 1956-60, Mufulira Branch Chairman, 1960-63, AMWU President. UNIP MP, 1964-71. Minister of State, Ministry of Commerce & Industry.

Edward Kabundi
Male. *Year of birth:* 1948. *Place of birth:* Mpika, NP.
Year of employment: 1968. *Place(s) of work:* Konkola mine.
Jobs: Clerk, Manning Centre; Safety Inspector; Clerk, Planning Dept. Retired 1999.
MUZ positions: 1985-88, Shop Steward.

Maxwell Kalesha-Chisoko

Male. *Year of birth:* 1935. *Place of birth:* Kawambwa, LP. *Year of employment:* 1956. *Place(s) of work:* Nchanga mine and townships. *Jobs:* Watchman; Interpreter; Whistleman; Loco Driver; Blasting Licence Holder; Skip Man; Tramming Section Boss; Asst Foreman; Township Inspector. Retired 1994.

ZMU/MUZ positions: 1965, Shop Steward; 1966, Chief Shop Steward; 1968, Nchanga Branch Vice Chairman; 1970-82, Branch Chairman. 1978-94, National Trustee.

Anton Kaluba

Male. *Year of birth:* 1947. *Place of birth:* Isoka, NP. *Year of employment:* 1970.

Place(s) of work: Luanshya mine. *Jobs:* Workman; Blasting Licence Holder; Heavy Equipment Driver; Section Boss, Tramming.

MUZ positions: 1973, Shop Steward. 1990, Vice Treasurer, Luanshya-Mpatamatu sub-Branch. 1994-98, sub-Branch Chairman, Luanshya-Mpatamatu.

Alex Kalunga

Male. *Year of birth:* 1937. *Place of birth:* Kabwe, Central Province (CP).

Year of employment: 1965. *Place(s) of work:* Luanshya mine.

Jobs: Operator, Senior Operator, Engineering Department; Engineer, Power Plant. Resigned 1977, owns and runs bar in Mpatamatu to date.

ZMU/MUZ positions: 1966, Shop Steward. 1968, Chief Shop Steward. 1975, sub-Branch Chairman Luanshya-Mpatamatu.

Sam Kangwa

Male. *Year of birth:* Unknown. *Place of birth:* Unknown. *Year of employment:* 1971.

Place(s) of work: Mufulira mine. Left mine employment in 1986.

Jobs: Staff Learner; Section Boss; Shift Boss; Mine Captain.

No MUZ positions.

Patrick Kanyanta

Male. *Year of birth:* 1958. *Place of birth:* Unknown. *Year of employment:* 1986.

Place(s) of work: Nkana Central Services Division. Retrenched 2000.

Jobs: Various clerical positions.

MUZ positions: 1991, Shop Steward. 1992, Vice Treasurer, Nkana Main Branch. 1994-2000, Branch Vice Chairman. 1998, Acting Branch Chairman.

John Kapansa

Male. *Year of birth:* 1942. *Place of birth:* Mansa, LP.

Year of employment: 1964. *Place(s) of work:* Nchanga mine.

Jobs: Workman; Operator; Section Boss; Foreman, Metallurgical Dept. Retired 1988.

MUZ positions: 1968, Shop Steward. 197?, Chief Shop Steward. 1978, Nchanga Branch Vice Secretary. c.1982-88, Branch Secretary.

Robert Kapenda

Male. *Year of birth:* 1952. *Place of birth:* Kitwe, CBP.

Year of employment: 1975. *Place(s) of work:* Luanshya Mine Police; Personnel Department, Luanshya mine townships; Ndola Copper Refinery.

Jobs: Mine Policeman; Check Control Clerk; Township Officer; Personnel Officer. 1978-82: Works Councillor, Luanshya. Retrenched 1992.

MUZ positions: 1982-89, Luanshya Main Branch Chairman.

Raphael Kasonde

Male. *Year of birth:* 1941. *Place of birth:* Kasama, NP.

Year of employment: 1965. *Place(s) of work:* NCR; Luanshya mine.

Jobs: Tank House Worker; Helper; Welder. Retired 1998.

MUZ positions: 1967, Shop Steward. 1974, Luanshya Branch Vice Treasurer.

Boniface Kumwenda

Male. *Year of birth:* 1953. *Place of birth:* Unknown area, EP.

Year of employment: 1981. *Place(s) of work:* Nkana.

Jobs: Mine policeman; Engineering Section. Retrenched 2002.

MUZ positions:; Shop Steward late 1980s, Chief Shop Steward 1990-91. Held branch positions 1990s.

Basilio Kunda

Male. *Year of birth:* 1933. *Place of birth:* Luwingu, NP.

Year of employment: 1952. *Place(s) of work:* Luanshya mine, Nchanga mine, Open Pit.

Jobs: Lasher; Driller; Blasting Licence Holder; Tramming Supervisor; Ganger; Open Pit Driver; Tavern Supervisor; Parks and Gardens Supervisor. Retired 1989.

ZMU/MUZ positions: 1964, ZMU Committee member, Nchanga. 1966, Shop Steward. 1970-88, Nchanga Branch Treasurer.

Joseph Kunda

Male. *Year of birth:* 1951. *Place of birth:* Mpika, NP.

Year of employment: 1971. *Place(s) of work:* Nchanga mine.

Jobs: Lasher; Track Layer; Pump Sinker, Engineering.

MUZ positions: 1980–2001, Shop Steward.

George Lombe

Male. *Year of birth:* 1956. *Place of birth:* Luanshya, CBP.
Year of employment: 1976. *Place(s) of work:* Mufulira mine.
Jobs: Improver, Instrumentation; Artisan, Metal Fabrication; Computer Programmer.
MUZ positions: 1986, Shop Steward. 1988, Chief Shop Steward. 1994-98, Mufulira Branch Vice Chairman. 2002–06, Branch Chairman.

David Longwani

Male. *Year of birth:* 1944. *Place of birth:* Luanshya, CBP.
Year of employment: 1967, AMCO Construction Company, joined NCCM 1976.
Place(s) of work: Konkola mine. *Jobs:* Lasher; Ventilation Helper; Machine Operator; Box Timberman; Cementation Helper; Track Layer. Early retirement 1999.
MUZ positions: 1981, Shop Steward. 1985-99, Chief Shop Steward.

Peter Lusaka

Male. *Year of birth:* 1961. *Place of birth:* Lusaka.
Year of employment: 1984. *Place(s) of work:* Konkola mine.
Jobs: Loco worker; Mine Garage worker; Computer worker. Medical discharge 2003.
MUZ positions: 1988, Shop Steward. 1992, Chief Shop Steward. 1997 - 1998, Konkola Branch Vice Secretary.

Rodah Matoba

Female. *Year of birth:* 1958. *Place of birth:* Kawambwa, LP.
Year of employment: 1979. *Place(s) of work:* Luanshya mine and hospital.
Jobs: Stock Control Clerk; Senior Stock Control Clerk; Community Services, Assistant Accountant. Left employment 2001.
MUZ positions: 1987, Chief Shop Steward. 1991, Luanshya Main Branch Vice Treasurer. 1996 - 2000, Branch Vice Chairman.

Timothy Mfula

Male. *Year of birth:* 1968. *Place of birth:* Chipata, EP.
Year of employment: 1990. *Place(s) of work:* Nkana Mine; Mufulira Trades School.
Jobs: Craft Improver; Trades School Instructor.
MUZ positions: 1992, Shop Steward. 1990s, Nkana Branch Executive member.
1998-02: National Trustee.

Charity M'tonga

Female. *Year of birth:* 1948. *Place of birth:* Mufulira, CBP.

Year of employment: 1971. *Place(s) of work:* Nchanga mine and townships.

Jobs: Filing Clerk; Clerk; Social Worker, Community Services; Clerk, Senior Clerk, Maintenance Section. Retrenched 1999.

MUZ positions: 1978, Shop Steward. 1980, Chief Shop Steward. 1982, Acting Branch Vice Secretary, Nchanga. 1984-90, Branch Secretary. 1990-99, Chief Shop Steward. 1999–date, MUZ Organising Secretary, Nchanga.

George Mukoma

Male. *Year of birth:* 1956. *Place of birth:* Lusaka. *Year of employment:* 1988.

Place(s) of work: Konkola. *Jobs:* Ventilation Officer. Left employment 2001.

MUZ positions: 1984, Shop Steward. 1990-95, Konkola Branch Chairman.

Henry Mukuka

Male. *Year of birth:* 1946. *Place of birth:* Ndola, CBP.

Year of employment: 1965. *Place(s) of work:* Luanshya.

Jobs: Lasher; Blasting Licence Holder; Timber Hangman; Section Boss, Production. Retrenched 1999.

MUZ positions: 1998-99, Luanshya-Mpatamatu sub-Branch Chairman.

Steven Mukuka

Male. *Year of birth:* 1946. *Place of birth:* Kasama, NP.

Year of employment: 1970. *Place(s) of work:* Nkana mine and townships.

Jobs: Lasher; Mechanic; Social Case Worker.

MUZ positions: 1972-74, Shop Steward. 1988, Chief Shop Steward. 1989–2000?, Nkana Main Branch Chairman. Resigned 2001.

2001–date, Wusakile-Chamboli MMD MP. 2002-03, Deputy Minister, Office of the Vice President.

Joseph Mulenga

Male. *Year of birth:* 1952. *Place of birth:* Unknown.

Year of employment: 1978. *Place(s) of work:* Nchanga Mine and Open Pit.

Jobs: Staff Learner; Driver/Operator; acting Shift Boss. Retrenched 1992.

MUZ positions: 1979–84, Shop Steward.

George Mulenga
Male. *Year of birth:* 1958. *Place of birth:* Mufulira, CBP.
Year of employment: 1975. *Place(s) of work:* Mufulira mine.
Jobs: Electrical Mechanic; Underground Maintenance; Accountant, Finance Department. Retrenched 1999.
MUZ positions: 1978, Shop Steward. 1982, Chief Shop Steward. 1982, Mufulira Branch Vice Secretary. 1985, Chief Shop Steward. 1987, Branch Vice Secretary. 1990, Branch Secretary. 1997-98, National Trustee.
1999–date, MUZ Organising Secretary, Luanshya.

Sefelino Mumba
Year of birth: 1949. *Place of birth:* Unknown area, LP.
Year of employment: 1971. *Place(s) of work:* Nchanga mine, Open Pit, Konkola mine.
Jobs: Employment Clerk; Senior Mechanic; Diesel Motor Mechanic; Leading Mechanic – Fitter; Senior Mechanic – Fitter. Dismissed in 1990. Reinstated in 1992. Retired 1999.
MUZ positions: 1975, Shop Steward. 1986-90, Chief Shop Steward.

Oswell Munyenyembe
Male. *Year of birth:* 1954. *Place of birth:* Isoka, NP.
Year of employment: 1976. *Place(s) of work:* Mufulira mine.
Jobs: Underground Workman; Survey Draughtsman; Stock Planner; Senior Draughtsman.
MUZ positions: 1984, Shop Steward. 1985, Mufulira-Butondo sub-Branch Vice Secretary. 1986-90, Mufulira Main Branch Secretary. 1998-2002, Mufulira Main Branch Chairman. 2002-date, National Vice President.

Abel Musonda
Male. *Year of birth:* 1932. *Place of birth:* Kawambwa, LP.
Year of employment: 1965. *Place(s) of work:* Luanshya Mine Hospital. Left mine employment 1970.
AMWU/ZMU/MUZ positions: 1955, Clerk, Bancroft AMWU. 1957, Organising Secretary, Bancroft.
1958, Organising Secretary, Luanshya. 1961, AMWU Assistant General Secretary.
1963-65, AMWU/ZMU President. 1965-66 and 1970, Luanshya Branch Chairman.

Evelyn Musonda
Female. *Year of birth:* 1955. *Place of birth:* Unknown.
Year of employment: 1974. *Place(s) of work:* Luanshya mine offices.
Jobs: Mailroom Clerk; Filing Clerk; various clerical posts in Payroll Department.
MUZ positions: 1982, Shop Steward. c.1985, Chief Shop Steward. 1990–2002, Luanshya Branch Secretary. 2002–date, National Trustee.

Stephen Musonda
Male. *Year of birth:* 1950. *Place of birth:* Mporokoso, NP.
Year of employment: 1971. *Place(s) of work:* Nchanga Open Pit.
Jobs: Trainee Mechanic; Senior Mechanic; Artisan Fitter. Medical discharge, 2001.
MUZ positions: 1978, Shop Steward. 1980-91, Chief Shop Steward. 1993, Nchanga Branch Vice Chairman. 1995-96, Nchanga Branch Chairman.

Fidelis Mwamba
Male. *Year of birth:* 1936. *Place of birth:* Unknown area, NP.
Year of employment: 1960. *Place(s) of work:* Mufulira, Luanshya mines.
Jobs: Carpenter; Assistant Foreman, Engineering Department. Retired late 1980s. Caretaker at MUZ Katilungu House.
MUZ positions: 1963-66, Chairman Mufulira Branch. 1967-75, Luanshya-Mpatamatu sub-Branch Chairman. 1974-78, National Treasurer.

Alfred Mwango
Male. *Year of birth:* 1950. *Place of birth:* Chilubi Island, LP. *Year of employment:* 1970. *Place(s) of work:* Luanshya mine. *Jobs:* Lasher; Trammer. Retrenched, 1999.
No MUZ positions.

Pascale Mwansa
Male. *Year of birth:* 1960. *Place of birth:* Mufulira, CBP. *Year of employment:* 1976. *Place(s) of work:* Mufulira mine. *Jobs:* Workman; Whistleman; Loco Driver; Blasting Licence Holder. Medical discharge 2002.
MUZ positions: 1985, Shop Steward. 1998-2002, Mufulira Main Branch Vice Chairman.

Raphael Mwanza
Year of birth: 1962. *Place of birth:* Petauke, EP.
Year of employment: 1985. *Place(s) of work:* Mufulira mine.
Jobs: Long-Haul Driller; Fireman; Leading Fireman; Fire Prevention Officer.
MUZ positions: 1990, Shop Steward. 1991-93, Chief Shop Steward.

Nellie Mweemba
Female. *Year of birth:* Unkown. *Place of birth:* Unknown.
Year of employment: Unknown. *Place(s) of work:* Nkana Mine hospitals. *Jobs:* Nurse.
MUZ positions: 1988, Shop Steward. 199?, Nkana Branch Treasurer.

David Ndalama
Male. *Year of birth:* 1947. *Place of birth:* Kasama, NP.
Year of employment: 1971. *Place(s) of work:* Nchanga mine, Open Pit.
Jobs: Heavy Equipment Driver; Safety Inspector; Heavy Equipment Training Inspector; Production Shift Boss; Mine Captain Training. Retired 2002.
MUZ positions: 1977, Shop Steward. 1983-88, Chief Shop Steward.

Agnes N'gandwe
Female. *Year of birth:* 1958. *Place of birth:* Kitwe, CBP. *Year of employment:* 1978. *Place(s) of work:* Mufulira hospital; primary school. *Jobs:* Nurse, School Nurse.
MUZ positions: 1988, Shop Steward. 1988, Branch Vice Treasurer. 1995, Acting Branch Treasurer. 1996–2002, Branch Treasurer.
1994–2000, Member, MUZ National Womens' Committee.

B J Ngulube
Male. *Year of birth:* 1954. *Place of birth:* Ndola, CBP. *Year of employment:* 1980. *Places of work:* Nkana mine. *Jobs:* Lasher; Engineering Section; Fire Brigade.
MUZ positions: 1981, Shop Steward. 1983, Chief Shop Steward. 1985-99, Nkana Main Branch Treasurer.

Monica Ngulube
Female. *Year of birth:* 1953. *Place of birth:* Lusaka.
Year of employment: 1976. *Place(s) of work:* Hospital, clinics in Mufulira. *Jobs:* Nurse.
MUZ positions: 1990, Shop Steward. 1998, Branch Vice Treasurer. 2002-date, Branch Secretary.

A K Nguluwe
Male. *Year of birth:* 1954. *Place of birth:* Petauke, EP.
Year of employment: 1974. *Place(s) of work:* Nkana mine.
Jobs: Lasher; Fitter; Junior Mechanic; Senior Mechanic; Winding Engine Driver.
MUZ positions: 1982, Shop Steward. 1986, Chief Shop Steward. 1990, Branch Treasurer. 1994, Chairman, Mindolo sub-Branch to date.

Harrison Nkonde
Male. *Year of birth:* 1956. *Place of birth:* Mufulira, CBP.
Year of employment: 1977. *Place(s) of work:* Chibuluma, Chambishi, Mufulira mines.
Jobs: Workman; Blasting Licence Holder; Tramming Supervisor; Tramming Section Boss; Shaft Timberman; Safety Inspector; Safety Officer.
MUZ positions: 1985, Shop Steward. 1986, Chief Shop Steward. 1992, Mufulira-Butondo sub-Branch Vice Chairman. 1994-98, Butondo sub-Branch Chairman.

Borries Ntankula
Male. *Year of birth:* 1960. *Place of birth:* Luanshya, CBP.
Year of employment: 1985. *Place(s) of work:* Nkana.
Jobs: Staff Learner; Housing Clerk; Laboratory Assayer.
MUZ positions: 1987, Shop Steward. 198?, Chief Shop Steward. 199?, Vice Treasurer. 1993, Branch Vice Secretary. 1995–2000, Branch Secretary.

Fenwick Nyirongo
Male. *Year of birth:* 1952. *Place of birth:* Isoka, NP.
Year of employment: 1976. *Place(s) of work:* Nchanga mine.
Jobs: Workman; Machine Operator; Crew Boss; Section Boss, Timber Crew.
MUZ positions: 1982 – date, Shop Steward.

Kramer Nyondo
Male. *Year of birth:* 1948. *Place of birth:* Unknown.
Year of employment: 1967. *Place(s) of work:* Nkana Smelter, Acid Plant.
Jobs: Trainee Operator; Operator; Foreman; Assistant General Foreman.
MUZ positions: 1969-early 1990s, Shop Steward.

Mary Palesi
Female. *Year of birth:* 1956. *Place of birth:* Mufulira, CBP.
Year of employment: 1978. *Place(s) of work:* Mine Police, Medical Department.
Jobs: Policewoman; Receptionist; Senior Clerk. Dismissed in 1992.
MUZ positions: 1979, Shop Steward. 1982, Chief Shop Steward. 1986-91, Mufulira Branch Treasurer. 1990s-date, MUZ Mufulira Organising Secretary.

Cameron Pwele
Male. *Year of birth:* 1958. *Place of birth:* Kitwe, CBP.
Year of employment: 1976. *Place(s) of work:* Nchanga, Kabwe, Luanshya.
Jobs: Laboratory Technician. Retrenched 2000.
MUZ positions: 1982-88, Chief Shop Steward, Nchanga. 1989-94, Shop Steward, Kabwe. 1998–2000, Branch Chairman, Luanshya Main.
2001-05: Luanshya-Roan UNIP MP (died 2005).

Robson Salimo
Male. *Year of birth:* Unknown. *Place of Birth:* Unknown.
Full-time ZMU/MUZ employee, 1964–date. Clerk, Organising Secretary at various branches. Administrative employment at MUZ Head Office.

Japhet Sekwila
Male. *Year of birth:* 1957. *Place of birth:* Luanshya, CBP.
Year of employment: 1976. *Place(s) of work:* Mufulira mine.
Jobs: Whistleman; Machine Operator; Engineer; Artisan.
MUZ positions: 1990, Shop Steward. 1990, Chief Shop Steward. 1991, Mufulira Branch Vice Chairman.

Timothy Walamba

Male. *Year of birth:* 1938. *Place of birth:* Kawambwa, LP.
Year of employment: 1959. *Place(s) of work:* Bancroft (Konkola) mine.
Jobs: Lasher; Concentrator worker; Track Layer; Timberman; Mechanic; Winding Engine Driver. Full-time MUZ official, 1982 – 1990. Retired 1990.
AMWU/ZMU/MUZ positions: 1964, Bancroft Branch Secretary. 1966, Branch Vice Chairman. 1971, Branch Chairman. 1978, National Vice Chairman. 1982-90, National Chairman.

Roy Yumbe

Male. *Year of birth:* 1932. *Place of birth:* Ndola, CBP. *Year of employment:* 1959.
Place(s) of work: Nkana mine. Jobs: Clerk; Winding Engine Driver. Retired 1984.
MASA positions: 1960, Shop Steward. 196?, Branch Treasurer.
MUZ positions: 1966 - 1967, Nkana Branch Secretary. 1970, Branch Vice Chairman. 1971. 1973, Branch Secretary. 1973-84, National Trustee. Branch Chairman, dates unclear.

APPENDIX 2: ZAMBIAN COPPER MINING INDUSTRY STATISTICS

	LME copper price (per Tonne)	Copper production (Tonnes)	Mineral sales revenue	Employment	Industry strike hrs
1965	£469	N/A	£177m	48,000	N/A
1966	£550	646,000	£217m	49,000	N/A
1967	£418	682,000	$644m	50,000	N/A
1968	£525	727,000	$745m	50,000	N/A
1969	N/A	N/A	N/A	N/A	N/A
1970	£589	683,000	$963m	N/A	N/A
1971	£444	633,000	$682m	49,748	11,000
1972	£428	698,122	$760m	N/A	13,000
1973	£481	N/A	$1184m	52,792	920
1974	£878	709,480	$1374m	56,128	4,774
1975	£557	647,163	$774m	N/A	54,529
1976	£781	711,681	$554m	N/A	4,351
1977	£750	658,788	$561m	59,055	8,128
1978	£710	654,012	$541m	N/A	4,152

	LME copper price (per Tonne)	Copper production (Tonnes)	Mineral sales revenue	Employment	Industry strike hrs
1979	£936	583,894	$958m	N/A	21,890
1980	£530	609,935	$818m	57,743	154,865
1981	£952	587,918	$895m	N/A	2,299,419
1982	£922	591,853	$893m	N/A	101
1983	£896	575,518	$915m	N/A	0
1984	£1,037	551,021	$1,522m	N/A	5,382
1985	£1,091	525,811	$791m	63,529/53,099	5,109,824
1986	£1,046	463,354	$594m	N/A	N/A
1987	£918	470,982	$788m	60,477	N/A
1988	£1,192	473,084	$1,485m	N/A	N/A
1989	£1,585	415,645	$1,741m	65,313	N/A
1990	£1,639	448,468	$1,282m	67,448	N/A
1991	£1,442	421,590	$1,246m	68,367	N/A

For 1985, first figure is total ZCCM employees. Second figure is employees directly employed in mining activities, indicative of diversification of ZCCM activities from mid 1980s. Subsequent figures (1986 - 1991) are total ZCCM employees.

Sources: ZCCM Mining Year Books, 1985 - 1991. Zambia Mining Year Books, 1965 - 1986; NCCM Mining Year Books, 1982 - 1984.

BIBLIOGRAPHY

UNPUBLISHED PRIMARY SOURCES

National Archives of Zambia

NAZ Reference	*Title*	*Dates covered*
LSS1/15/277	Department of Labour Monthly Report	1964 - 1966
LSS1/15/297	Monthly reports, Ndola	1964 - 1968
LSS1/17/70	Industrial Relations Bill Law Reform	1969 - 1970
LSS1/19/220	UTUC	1964 - 1965
LSS1/19/224	Northern Rhodesian AMTWU	1964 - 1966
LSS1/19/226	UTUC soon to be ZCTU	1965 - 1966
LSS1/19/227	ZCTU	1965 - 1967
LSS1/19/228	ZMU	1965 - 1967
LSS1/19/238	ZCTU	1966 - 1968
LSS1/19/243	ZCTU	1967 - 1968
LSS1/19/245	MUZ	1967 - 1969
LSS1/19/260	ZCTU	1969
LSS1/19/269	MUZ	1971
LSS1/20/89	TUs & Trade Disputes Ordinance (Amendment) Act, Legal Drafting	1964
LSS1/20/96	ZCTU	1971
unnumbered	MoL Tripartite Consultative Council	1988
LSS/102/1/17	Luanshya Annual Reports	1969– 1995
HM77	Papers of Peter Matoka	
HM89	Papers of Alex Shapi	

ZCCM Archives

ZCCM Reference	*Title*	*Dates covered*
1.3.1H	ACME-MUZ Dispute	1979 - 1988
1.4.5B	MUZ Constitution	1967 - 1995
1.5.3G	Industrial Relations Act	1970 - 1995
1.7.1E	GIRMs papers	1989 - 1990
1.7.5C	African Labour Political	1964 - 1965
1.9.1D	CISB Personnel Research Unit report, 'Workers' Participation'	1972
1.9.2F	ZMU General	1965
1.9.2F	ZMU General	1966
1.9.4D	IRB, ACME-MJIC mtgs	1983 - 1987
3.1.4F	MUZ Elections etc	1970
3.1.4F	MUZ Negotiations	1973 - 1979
3.1.4F	MUZ	1974
3.1.4F	MUZ, general	1967 - 1970

ZCCM Reference	*Title*	*Dates Covered*
4.5.1C	MUZ Wage Negotiations	1979
4.5.1D	MUZ Negotiations	1981
4.5.2I	Work Stoppages, Strikes	1972 - 1978
6.6.2H	Luanshya Mine Police Unit reports	1990 - 1991
7.1.4I	Industrial Relations Bureau, MJIC, MUZ mtgs, minutes	1970 - 1974
8.6.1D	Tripartite Labour Consultative Committee Labour Dept reports	1989 - 2000
10.6.10F	Monthly Comments on Industrial Relations	1963 - 1965
11.6.2A	Conciliations, arbitrations, strikes, disputes	1965
11.6.2A	Conciliations, Arbitrations, Strikes, Disputes	1966 - 1968
11.6.2A	Conciliations, Arbitrations, Strikes, Disputes	1968 - 1970
13.4.10E	MJIC/MUZ/ACME	1975 - 1980
15.2.1C	Government Takeover	1969 - 1971
15.2.1C	Government, General Political	1973 - 1978
15.2.6B	Ministry of Mines and Industry, Annual Reports and Correspondence	1972 - 1978
16.2.8F	Work Stoppage ZMU Local Disputes, Bancroft	1965 - 1967
16.2.9B	GIRMS Job Evaluation Ctte Mtgs and related papers	1973 - 1974
16.2.9B	MJIC-MUZ-ACME mtgs	1974 - 1975
16.2.9C	GIRM Ctte, related correspondence	1986
16.3.2B	MJIC-ZMU Negotiations	1965 - 1966
16.3.2C	MUZ/ACME	1990 - 1992
17.2.2C	Local Disputes, Work Stoppages, Chibuluma	1965 - 1968
17.2.3B	ZMU General	1966
17.2.3B	ZMU Local Disputes, NCR	1964 - 1967
19.2.11D	Meetings, Disputes & Union Affairs	1967
19.3.2C	Konkola Division, Work Stoppages, Strikes and Man Hours Lost	1966 - 1967
Unnumbered	Zambia Mining Year Books	1965 - 1986

Mineworkers' Union of Zambia (MUZ) records

N.B. Official titles and numbering for files is provided when given. Otherwise, a description of the contents of the file is provided.

MUZ Reference	*Title*	*Dates Covered*
MUZ Head Office (HO)		
IB/B/71	Other Govt Ministries and Govt Bodies	1968 - 1981

MUZ Head Office (HO) (cont.)

MUZ/5/8/73	MUZ Supreme Council minutes	1972 - 1976
MUZ/15/13/81	Other Govt Ministries and Govt Bodies	1981 - 2001
	MUZ Conferences, other documents	1967 - 1970
	CEC minutes and related papers	1986 - 1997
	MJIC meetings	1970
	MJIC papers	1985 - 1991
	Supreme Council & CEC minutes	1975 - 1981
	Supreme Council minutes	1981 - 1987
	Supreme Council minutes	1983 - 1988
	Annual conference files	1972 - 1976
	Research Dept file	1977 - 1989
	Supreme Council and CEC mins	1975 - 1981
	UNIP, Politics	1977 – 1989
	ZCTU-related papers	1977 - 1992

MUZ Luanshya

MUZ/5/A	Biannual [sic] Conference	1970 - 1984
MUZ/5/A	Supreme Council	1967 - 1968
MUZ/5/B	Supreme Council	1970 - 1973
MUZ/5/D	CISB Negotiating Committee – MJIC, Roan Branch	1967 - 1969
MUZ/6/A	Luanshya Joint Executive Committee and Joint Caucus Mtgs	1972 - 1981
MUZ/6/C	Mtgs with Local Management	1980 - 1983
MUZ/17/A	UNIP	1971 – 1974
	General Correspondence	1980 - 1985
	Mtgs with Management	1975 - 1978
	Mtgs with Management	1978 - 1980
	Branch mtgs with Management, Steering Ctte	1979 - 1981
	Supreme Council	1964 - 1982
	File of BVC H Chibwe	1963 - 1969
	Untitled, grievance cases, other correspondence	1970 - 1981

MUZ Mufulira

MUZ/5/A/2/90	Biennial Conference Corresp file	1990 - 1998

MUZ Nchanga

MUZ/5/B	Supreme Council	1979 - 1985
MUZ/6/A	Branch Executive correspondence	1980 - 1987
MUZ/11/B	Public Meetings	1972 - 1986
MUZ/17/A	CISB Correspondence	1987 - 1991

MUZ Nkana

MUZ/11/B	Public meetings and regional activities	1967 - 1990
	Branch Officials mtgs with management	1976 - 1990
	Complaint files	1985 - 1987
	SC and CEC meetings	1970 - 1989
	Branch Executive Committee meetings	1990 - 1991
	ZCTU and Labour Day	1967 - 1994

PRINTED PRIMARY SOURCES

Newspapers

Northern News (Ndola, Zambia) 1964

Times of Zambia (Ndola, Zambia) 1965 - 1991

Other Primary Sources

Economic Development Institute of the World Bank/R. Gulhati, *Impasse in Zambia – The Economics and Politics of Reform* Development Policy Case Series No.2 (Washington, 1989)

International Bank for Reconstruction and Development, *Wage Policy and the Structure of Wages and Employment in Zambia* Discussion Paper (Washington, January 1986)

International Labour Organisation/United Nations Development Programme /Technical Advisory Service, *Report to the Government of Zambia on Incomes, Wages & Prices in Zambia: Policy & Machinery* (Geneva, 1969)

Inter-African Network for Human Rights & Development (Afronet), Citizens for a Better Environment (CBE) & Rights and Accountability in Development (RAID), *Zambia: Deregulation and the Denial of Human Rights* Submission to the OECD Committee on Economic, Social and Cultural Rights (Lusaka, Kitwe, Oxford, March 2000)

M.F. McPherson, *Wage Leadership and Zambia's Mining Sector – Some Evidence* Harvard Institute for International Development Discussion Paper No.43 (Cambridge MA, October 1978)

E. Sibanda, *Workers' Participation, Collective Bargaining and Tripartism in Formation of Employment Policies in Tanzania, Zambia and Zimbabwe* Zimbabwe Institute of Development Studies Discussion Papers, No.8 (Harare, 1991)

Transparency International Zambia, *Transparency and The Rule of Law in The Privatization of The Zambia Consolidated Copper Mines (ZCCM) Assets* (Lusaka, 2002)

H.A. Turner, *Wage Trends, Wage Policies and Collective Bargaining: the Problems for Underdeveloped Countries* University of Cambridge Department of Applied Economies Occasional Papers, No.6 (Cambridge, 1965)

United Nations/Economic Commission for Africa/Food & Agriculture Organisation, *Economic Survey Mission on the Economic Development of Zambia* (Ndola, 1964)

United Nations Development Programme/International Labour Organisation, *Second Report to the Government of Zambia on Incomes, Wages & Prices in Zambia: Policy & Machinery* (Geneva, 1978)

United Nations Industrial Development Organisation, *The Republic of Zambia* Industrial Review Series (New York, February 1985)

World Bank Country Economics Department WPS 222/C. Colclough, *The Labour Market & Economic Stabilisation in Zambia* (Washington, 1989)

Zambia, *Kenneth Kaunda – on the Redemption of ZIMCO Bonds, 31st August 1973* (Lusaka, 1973)

Zambia, *Kenneth Kaunda – The Dignity of Labour, speech 29th April 1972* (Lusaka, 1972)

Zambia, *Report into the Commission of Inquiry into the Mining Industry of Zambia* (Lusaka, 1966)

Zambia, *Report of the National Commission on the Establishment of a One-Party Participatory Democracy in Zambia* (Lusaka, 1972)

Zambia, *Third National Development Plan 1979-1983* (Lusaka, 1979)

Zambia, *Towards Self-Sufficiency Through Development* (Lusaka, 1971)

Zambia, *Zambia's Guidelines for the Next Decade* (Lusaka, 1968)

Zambia, *Zimco – Core of the Zambian Economy* (Lusaka, 1984?)

Zambia Congress of Trade Unions and International Confederation of Free Trade Unions, R.S.K. Muchimba (ed.), *National Seminar on Employment* (Lusaka, 1979)

Zambia Congress of Trade Unions/R.S. Muchimba, *Trade Unionism in Zambia* ZCTU (Kitwe, 1980?)

Zambia Department of Labour, *Annual Reports 1964 - 1986* (Lusaka, various)

Zambia – European Community, *Country Strategy Paper and National Indicative Programme for 2001 – 2007* (Lusaka, 2001)

Zambia, Ministry of Development Planning & National Guidance, *Second National Development Plan January 1972 – December 1976* (Lusaka, 1971)

Zambia Ministry of Labour, *Livingstone Labour Conference, 7th-8th April 1967* Government Printers (Lusaka, 1967)

Zambia Office of the Prime Minister, Provincial & Local Government Administration Division, *Copperbelt Province – Annual Report* (Lusaka, various)

Zambia Office of the Prime Minister, Department of Industrial Participatory Democracy, *Report & Recommendations of the National Symposium on Industrial Participatory Democracy in Zambia, 12th– 2nd July 1976* (Lusaka, 1977)

Zambian High Commission, *President Bans UPP* (London, 1972)

SECONDARY SOURCES

Books

R. Abrahamsen, *Disciplining Democracy: Development Discourse and Good Government in Africa* Zed Books (London, 2000)

E.O. Akwety, *Trade Unions and Democratisation: a Comparative Study of Zambia and Ghana* University of Stockholm, Department of Political Science (Stockholm, 1994)

D.J. Alexander, *The Development of Workers' Education and Political Change in Zambia* Occasional Papers No.42, Centre of African Studies, University of Edinburgh (Edinburgh, 1993)

G. Arrighi & J.S. Saul, *Essays on the Political Economy of Africa* Monthly Review Press (New York & London, 1973)

R.H. Bates, *Party-Union Relations in the Copperbelt of Zambia* Yale University Press (Cambridge MA, 1969)

R.H. Bates, *Rural Responses to Industrialization: a Study of Village Zambia* Yale University Press (New Haven CN, 1976)

R.H. Bates, *Unions, Parties, and Political Development: a Study of Mineworkers in Zambia* Yale University Press (New Haven CN & London, 1971)

B. Beckman & L. Sachikonye (eds.) *Labour Regimes and Liberalization: the Restructuring of State-Society Relations in Africa* University of Zimbabwe Publications (Harare, 2001)

E.L. Berger, *Labour, Race and Colonial Rule* Clarendon Press (Oxford, 1974)

M. Bratton, *The Local Politics of Rural Development: Peasant and Party-State in Zambia* University Press of New England (Hanover NH & London, 1980)

C. Brown, *'We Were All Slaves': African Miners, Culture, and Resistance at the Enugu Government Colliery* Heinemann (Portsmouth NH, 2003)

M. Burawoy, *The Colour of Class on the Copper Mines* UNZA, Institute of African Studies (Lusaka, 1972)

M. Burawoy, *The Politics of Production* Verso (London, 1985)

A. Campbell, *The Scottish Miners, 1874 – 1939, Vols 1 & 2* Ashgate (Aldershot, 2000)

S. Chan, (ed. C. Clancy), *Zambia and the Decline of Kaunda, 1994 – 1998* Edwin Mellen Press African Studies Volume 57 (Lewiston NY, 2000)

A.P. Cheater, *The Politics of Factory Organisation* Mambo Press (Gweru, 1986)

S.N. Chipungu (ed.), *Guardians in their Time: Experiences of Zambians under Colonial Rule, 1890 – 1964* Macmillan (London & Basingstoke, 1992)

B.S. Chisala, *The Downfall of President Kaunda* Co-op Printing (Lusaka, 1994)

R. Cohen, *Labour and Politics in Nigeria* Heinemann (London, 1974, 1981)

F. Cooper, *Decolonization and African Society: the Labour Question in French and British Africa* Cambridge University Press (Cambridge, 1996)

J. Crisp, *The Story of an African Working Class: Ghanaian Miners Struggle, 1870 – 1980* Zed Books (London, 1984)

P. Daniel, *Africanisation, Nationalisation and Inequality: Mining Labour and the Copperbelt in Zambian Development* Cambridge University Press (Cambridge, 1979)

T. Dunbar Moodie, with V. Ndatshe, *Going for Gold: Men, Mines & Migration* University of California Press (Berkeley CA, 1994)

A.L. Epstein, *Politics in an Urban African Community* Institute for African Studies, University of Zambia/Manchester University Press (Manchester, 1958, 1973)

F. Fanon, *The Wretched of the Earth* MacGibbon & Kee (London, 1965)

J. Ferguson, *Expectations of Modernity: Myths and Meanings of Modern Life on the Zambian Copperbelt* University of California Press (Berkeley CA, 1999)

B. Freund, *Capital and Labour in the Nigerian Tin Mines* Longman (Harlow, 1981)

B. Freund, *The African Worker* Cambridge University Press (Cambridge, 1988)

C. Gertzel (ed.), C. Baylies, & M. Szeftel, *The Dynamics of the One-Party State in Zambia* Manchester University Press (Manchester, 1984)

R. Gray, *The Two Nations – Aspects of the Development of Race Relations in the Rhodesias and Nyasaland* Oxford University Press (Oxford, 1960)

R. Hall, *Zambia, 1890-1964: The Colonial Period* Longman (London 1965, 1976)

M. Hamalengwa, *Class Struggles in Zambia, 1889 - 1989, and the Fall of Kenneth Kaunda, 1990 - 1991* University Press of America (Lanham MA, 1992)

P. Harries-Jones, *Freedom and Labour: Mobilization and Political Control on the Zambian Copperbelt* Basil Blackwell (Oxford, 1975)

C. Harvey (ed.), *Constraints on the Success of Structural Adjustment Policies in Africa* Macmillan (London 1996)

J.A. Higginson, *A Working-Class in the Making: Belgian Colonial Labor Policy, Private Enterprise and the African Mineworker, 1907 - 1951* University of Wisconsin Press (Madison WI, 1989)

R. Hyman, *Marxism and the Sociology of Trade Unionism* Pluto Press (London, 1971, 1973)

International Marxist Group, *Zambia: Humanist Rhetoric, Capitalist Reality* Africa in Struggle Occasional Papers No.2 (London, 1975)

W.R. Jacobs, *The Relationship Between African Trade Unions and Political Organisations in Northern Rhodesia/Zambia, 1949-1964* International Institute for Labour Studies (Geneva, 1971)

V. Jamal & J. Weeks, *Africa Misunderstood: or Whatever Happened to the Rural-Urban Gap?* International Labour Organization, Macmillan (Basingstoke & London, 1993)

R. Jeffries, *Class, Power & Ideology in Ghana: the Railwaymen of Sekondi* Cambridge University Press (Cambridge, 1978)

F.A. Johnstone, *Class, Race and Gold: a Study of Class Relations and Racial Discrimination in South Africa*, Routledge & Kegan Paul (London, 1976)

N. Kamwambe, *Frederick Chiluba: Is he Riding a Tide of Fortune?* Shelley's Printers (Lusaka, 1991)

B. Kapferer, *Strategy and Transaction in an African Factory* Manchester University Press Imprint (Manchester, 1972)

F. Kaunda, *Selling the Family Silver: the Zambian Copper Mines Story* Printpak Books (KwaZulu Natal, 2002)

K.D. Kaunda, *Humanism in Zambia and a Guide to its Implementation* Zambia Information Services (Lusaka, 1968)

K.D. Kaunda, *Humanism in Zambia and a Guide to its Implementation Part II* Zambia Information Services (Lusaka, 1974)

K.D. Kaunda, *Towards Complete Independence* Zambia Information Services (Lusaka, 1969)

T.M. Klubock, *Contested Communities: Class, Gender, and Politics in Chile's El Teniente Copper Mine, 1904 - 1951* Duke University Press (Durham NC, 1998)

P.M. Lubeck, *Islam and Urban Labor in Northern Nigeria: the Making of a Muslim Working-Class* Cambridge University Press (Cambridge, 1986, 1988)

J. Lunn, *Capital and Labour on the Rhodesian Railway System, 1888 - 1947* Macmillan (Basingstoke, 1997)

P.D. Machungwa & T.K. Mwaba, *Works Councils and Other Committees: Aspects of Industrial Relations* M & M Management and Labour Consultants (Lusaka, 1989)

A. Martin, *Minding Their Own Business: Zambia's Struggle Against Foreign Capital* Hutchinson (London, 1972)

A. Mbikusita–Lewanika, *Milk in a Basket! – The Political-Economic Malaise in Zambia* Zambia Research Foundation (Lusaka, 1990)

H.S. Meebelo, *Reaction to Colonialism* Institute of African Studies University of Zambia/Manchester University Press (Manchester, 1971)

H.S. Meebelo, *African Proletarians and Colonial Capitalism* Kenneth Kaunda Foundation (Lusaka, 1986)

R. Michels, *Political Parties: a Sociological Study of the Oligarchical Tendencies of Modern Democracy* Glencoe (1911)

J.C. Mitchell, *Cities, Society & Social Perception: a Central African Perspective* Clarendon Press (Oxford, 1987)

H.L. Moore & M. Vaughan, *Cutting Down Trees: Gender, Nutrition and Agricultural Change in the Northern Province of Zambia, 1890 - 1990* Heinemann/James Currey/UNZA Press (Portsmouth NH, London, Lusaka 1994)

E. Mudenda, *A Generation of Struggle* SAPES Books (Harare, 199?)

D.C. Mulford, *Zambia: The Politics of Independence 1957 - 1964* Oxford University Press (Oxford, 1967)

J. Mwanakatwe, *Teacher, Politician, Lawyer: My Autobiography* Bookworld Publishers (Lusaka, 2003)

G.B. Mwangilwa, *The Kapwepwe Diaries* Multimedia Publications (Lusaka, 1986)

F.K. Mwanza, *Chasing the Winds and Dependency Syndrome: Topical Issues of Zambia's Political Economy* MFK Management Consultancy Services (Lusaka, 1993)

M.R. Mwendapole, *A History of the Trade Union Movement in Zambia up to 1968* UNZA Institute of African Studies, Communication No.13 (Lusaka, 1977)

M.B.N. Nguvu, *The Role of Trade Unions and Employers Organisations in Socio-Economic Development and Employment Promotion in Zambia* International Institute for Labour Studies (Geneva, 1976)

P. Nordlund, *Organising the Political Agora: Domination and Democratisation in Zambia and Zimbabwe* Uppsala University (Uppsala, 1996)

P.O. Ohadike, *Development of and Factors in the Employment of African Migrants in the Copper Mines of Zambia, 1940 - 1966* University of Zambia (Lusaka, 1969)

P.E. Ollawa, *Participatory Democracy in Zambia* Stockwell (Ilfracombe, 1979)

K. Osei-Hwedie & M. Ndulo, *Issues In Zambian Development* Omenana Press (Nyangwe & Roxbury MA, 1985)

R. Palmer & N. Parsons (eds.) *The Roots of Rural Poverty in Central and Southern Africa* Heinemann (London, 1977)

J.L. Parpart, *Class Consciousness Among Zambian Copper Miners 1950 - 1966* African Studies Center, Boston University (Boston, 1982)

J.L. Parpart, *Labour and Capital on the African Copperbelt* Temple University Press (Philadelphia, 1983)

J.L. Parpart, *Working-Class Wives and Collective Labour Action on the North Rhodesian Copperbelt 1926 - 64* Pamphlet (Boston, 1982)

A.J. Peace, *Choice, Class & Conflict: a Study of Southern Nigerian Factory Workers* Harvester Press (Brighton, 1979)

C. Perrings, *Black Mineworkers in Central Africa* Heinemann (London, 1979)

I.R. Phimister & C. van Onselen, *Studies in the History of African Mine Labour in Colonial Zimbabwe* Mambo Press (Gwelo, 1976)

I.R. Phimister, *Wangi Kolia: Coal, Capital & Labour in Colonial Zimbabwe, 1894 - 1954* Baobab Books/Witwatersrand University Press (Harare & Johannesburg, 1994)

J. Pottier, *Migrants No More: Settlement and Survival in Mambwe Villages, Zambia* Indiana University Press (Bloomington IN & London, 1988)

H. Powdermaker, *Copper Town: Changing Africa* Harper (New York, 1962)

B. Raftopolous & I. Phimister, *Keep on Knocking, A History of the Labour Movement in Zimbabwe, 1900 - 97* Zimbabwe Congress of Trade Unions & Friedrich Ebert Stiftung/Baobab Books (Harare, 1997)

L. Rakner, *Trades Unions in Processes of Democratisation: a Study of Party-Labour Relations in Zambia* Michelsen Institute (Bergen, 1992)

L. Rakner, *Political and Economic Liberalisation in Zambia, 1991 - 2001* Nordiska Africainstitutet (Uppsala, 2003)

R. Reinikka-Soininen, *Theory and Practice in Structural Adjustment: the Case of Zambia* Helsinki School of Economics Research Paper (Helsinki, 1990)

A. Roberts, *A History of Zambia* Heinemann (London, 1976)

R.I. Rotberg, *The Rise of Nationalism in Central Africa: The Making of Malawi and Zambia, 1873 – 1964* Harvard University Press (Cambridge MA, 1966)

L. Sachikonye (ed.), *Democracy, Civil Society and the State: Social Movements in Southern Africa* SAPES Books (Harare, 1995)

R. Sandbrook & R. Cohen (eds.), *Development of an African Working-Class*
Longman (London, 1975)

A. Sardanis, *Africa: Another Side of the Coin* I B Tauris (New York, 2003)

F.B. Schiphorst, *Strengths & Weaknesses: The Rise of the Zimbabwe Congress of Trade Unions (ZCTU) and the Development of Labour Relations 1980-1995* Published PhD Thesis, University of Leiden (2001)

L. Schumaker, *Africanizing Anthropology: Fieldwork, Networks, and the Making of Cultural Knowledge in Central Africa* Duke University Press (Durham NC, 2001)

O. Sichone & B.C. Chikulo (eds.), *Democracy in Zambia: Challenges for the Third Republic* SAPES Books (Harare, 1996)

R. Sklar, *Corporate Power in an African State* University of California Press (Berkeley CA, 1975)

E.P. Thompson, *The Making of the English Working-Class* Penguin (London, 1963, 1968, 1980)

W. Tordoff (ed.), *Politics in Zambia* Manchester University Press (Manchester, 1984)

B. Turok (ed.), *Development in Zambia – A Reader* Zed Press (London, 1979)

B. Turok, *Mixed Economy in Focus: Zambia* Institute for African Alternatives (London, 1989)

G. Wilson & M. Wilson, *The Analysis of Social Change Based on Observations in Central Africa* Cambridge University Press (Cambridge, 1945)

K. Woldring (ed.), *Beyond Political Independence: Zambia's Development Predicament in the 1980s* Mouton Publishers (Berlin, 1984)

R. Young, *Zambia: Adjusting to Poverty* North-South Institute (Ottawa, 1988)

Zambia Congress of Trade Unions/International Confederation of Free Trade Unions-African Regional Organisation Research Project, H.C. Mpuku & A. Muneku (eds.), *The Impact of Structural Adjustment Programmes on the Zambian Economy: A Trade Union Perspective* Zambia Congress of Trade Unions (Kitwe, 2001)

Articles and Book Chapters

P. Alexander, 'Oscillating Migrants, 'Detribalised Families' and Militancy: Mozambicans on Witbank Collieries, 1918 – 1927', *Journal of Southern African Studies* 27, 3 (2001), pp. 505 - 525

D.M.C. Bartlett, 'Civil Society and Democracy: a Zambian Case Study', *Journal of Southern African Studies* 26, 3 (2000), pp. 429 - 446

R.H. Bates & P. Collier, 'The Politics and Economics of Policy Reform in Zambia', in R.H. Bates & A.O. Krueger (eds.), *Political and Economic Reform: Evidence from Eight Countries* Blackwell (Oxford and Cambridge MA, 1993)

C. Baylies, 'State and Class in Postcolonial Africa' in M. Zeitlin (ed.), *Political Power and Social Theory* Vol 5 JAI Press (Greenwich CN, 1985)

C. Baylies, 'Zambia's Economic Reforms and their Aftermath: The State and the Growth of Indigenous Capital', *Journal of Commonwealth and Comparative Politics* 20, 3 (1982), pp. 235 – 263

C. Baylies & M. Szeftel, 'The Rise of a Zambian Capitalist Class in the 1970s', *Journal of Southern African Studies* 8, 2 (1982), pp. 187 – 213

C. Baylies & M. Szeftel, 'Democratization and the 1991 Elections in Zambia', in J. Daniel, R. Southall & M. Szeftel (eds.), *Voting for Democracy: Watershed Elections in Contemporary Anglophone Africa* Ashgate (Aldershot, 1999)

E. Berg & J. Butler, 'Trade Unions', in J. Coleman, & C.G. Rosberg (eds.), *Political Parties and National Integration in Tropical Africa* University of California Press (Berkeley CA, 1964)

M. Bratton, 'Economic Crisis and Political Realignment in Zambia', in J. Widner, *Economic Change and Political Liberalization in sub-Saharan Africa* John Hopkins Press (Baltimore, 1994)

M. Burawoy, 'The Hidden Abode of Under-Development: Labour Process and the State in Zambia', *Politics and Society* 11, 2 (1982), pp. 123 – 166

F. Cooper, 'The Dialectics of Decolonization: Nationalism and Labour Movements in Postwar French Africa' in F. Cooper & A.L. Stoler (eds.), *Tensions of Empire: Colonial Cultures in a Bourgeois World* University of California Press (Berkeley CA, 1997)

J. Craig, 'Putting Privatisation into Practice: the case of Zambia Consolidated Copper Mines', *Journal of Modern African Studies* 39, 3 (2001), pp. 389 - 410

J. Ferguson, 'Modernist Narratives, Conventional Wisdoms and Colonial Liberalism: Reply to a Straw Man', *Journal of Southern African Studies* 20, 4 (1994), pp. 633 - 640

C. Gertzel, 'Labour and the State: The case of Zambia's Mineworkers' Union', *Journal of Commonwealth and Comparative Politics* 13, 3 (1975), pp. 290 – 304

C. Gertzel, 'Industrial Relations in Zambia to 1975', in U.G. Damachi, H.D. Seibel & L. Trachtman (eds.), *Industrial Relations in Africa* St. Martin's Press (New York, 1979)

P. Gibbon, 'Structural Adjustment and Pressures Towards Multi-Partyism in sub-Saharan Africa' in Y. Bangura, P. Gibbon & A. Ofstad (eds.) *Authoritarianism, Democracy & Adjustment: the Politics of Economic Reform in Africa* (Uppsala, 1992)

K. Good, 'Zambia and the Liberation of South Africa', *Journal of Modern African Studies* 25, 3 (1987), pp. 505 - 40

K. Good, 'Zambia: Back Into the Future', *Third World Quarterly* 10, 1 (1988), pp. 37 - 53

K. Good, 'Debt and the One-Party State in Zambia', *Journal of Modern African Studies* 27, 2 (1989) pp. 297 – 313

K.P. Good, 'The Reproduction of Weakness in the State and Agriculture: Zambian Experience', *African Affairs* 85, 339 (1986), pp. 239 - 265

H. Heisler, 'The Creation of a Stabilized Urban Society: a Turning Point in the Development of Northern Rhodesia/Zambia', *African Affairs* 70, 279 (1971), pp. 125 – 145

I. Henderson, 'Wage-Earners and Political Protest in Colonial Africa – the Case of the Copperbelt', *African Affairs* 72, 288 (1973), pp. 288 - 299

E.J. Hobsbawm, 'The Labour Aristocracy in 19th Century Britain', in *Labouring Men: Studies in the History of Labour* Weidenfeld & Nicholson (London, 1964)

S. Kibble, 'Zambia: Problems for the MMD', *Review of African Political Economy* 53 (1992), pp. 104 – 108

M. Larmer, "'If We Are Still Here Next Year': Zambian Historical Research in the Context of Decline', *History in Africa* 31 (New Brunswick NJ, 2004), pp. 215 - 229

M. Larmer, "'A Little Bit Like a Volcano': The United Progressive Party and resistance to one-party rule in Zambia, 1964 – 1980', *International Journal of African Historical Studies*, Vol. 39 (Boston, 2006)

M. Legassick, 'South Africa – Forced Labour, Industrialisation and Differentiation' in R. Harris (ed.), *The Political Economy of Africa* Schenkman (Cambridge MA, 1973)

R.T. Libby & M.E. Woakes, 'Nationalization and Displacement of Development Policy in Zambia', *African Studies Review* 13, 1 (1980), pp. 33 - 50

H. Macmillan, 'The Historiography of Transition on the Zambian Copperbelt – Another View', *Journal of Southern African Studies* 19, 4 (1993), pp. 681 – 712

G. Macola, '"It Means as if We Are Excluded from the Good Freedom": Thwarted Expectations of Independence and Royal Politics in the Luapula Province of Zambia, 1964-66', *Journal of African History*, 47, 1 (2006)

M. Martin, 'Neither Phoenix nor Icarus: Negotiating Economic Reform in Ghana and Zambia, 1983-92', in T.M. Callaghy & J. Ravenhill (eds.), *Hemmed In: Responses to Africa's Economic Decline* Columbia University Press (New York, 1993)

G. Mudenda, 'Class Formations and Class Struggle in Contemporary Zambia' in B. Magubane & Nzongola-Ntalaya (eds.), *Proletarianization and Class Struggle in Zambia* Contemporary Marxism No.6 (San Francisco, 1983)

F.B. Nyamnjoh, 'Expectations of Modernity in Africa or a Future in the Rear-view Mirror?', *Journal of Southern African Studies* 27, 2 (2001), pp. 363 - 369

M. Ogura, 'Rural-Urban Migration in Zambia and Migrants' Ties to Home Villages', *Developing Economies* 29, 2 (1991), pp. 145 - 165

F.O.E. Okafor, 'The Mining Multinationals and the Zambian Economy', *The African Review* 17, 1 & 2 (1990), pp. 66 - 79

J.L. Parpart, 'The Labour Aristocracy in Africa: The Copperbelt Case 1924-1967', *African Economic History* 13 (1984), pp. 171 - 191

J.L. Parpart & T.M. Shaw, 'Contradiction and Coalition: Class Fractions in Zambia, 1964-1984', *Africa Today* 30, 3 (1983), pp. 23 - 50

C. Perrings, 'A 'Moment in the Proletarianization' of the New Middle Class: Race, Value and the Division of Labour in the Copperbelt 1946 -1966', *Journal of Southern African Studies* 6, 2 (1980), pp. 183 - 213

C. Perrings, 'Premise and Inference in Labour Studies: a Zambian Example', *African Affairs* 81, 322 (1982), pp. 87 – 99

J. Petras, 'State Capitalism and the Third World', *Development and Change* 8, 1 (1977), pp. 1 - 17

B.J. Phiri, 'Zambia – Myths and Realities of One-Party Participatory Democracy', *Geneve-Afrique* 29, 2 (1991), pp. 9 - 24

D. Potts, 'Counter-Urbanization on the Zambian Copperbelt? Interpretations and Implications', *Urban Studies* 42, 4 (2005), pp. 583 – 609

D. Potts, 'Shall We Go Home? Increasing Urban Poverty in African Cities and Migration Processes', *Geographical Journal* 161, 3 (1995), pp. 245-264

D. Potts & C. Mutambirwa, 'Rural-Urban Linkages in Contemporary Harare: Why Migrants Need Their Land', *Journal of Southern African Studies* 16, 4 (1990), pp. 677- 698

R.L. Prain, 'The Stabilisation of Labour in the Rhodesian Copper Belt', *African Affairs* 55, 221 (1956), pp. 305 – 312

B. Raftopolous, 'The Labour Movement and the Emergence of Opposition Politics in Zimbabwe', in B. Raftopolous & L. Sachikonye, *Striking Back* Weaver Press (Harare & London, 2001)

D. Rothchild, 'Rural-Urban Inequalities and Resource Allocation in Zambia', *Journal of Commonwealth Political Studies* 10, 3 (1972), pp. 222 – 242

L. Sachikonye, 'State, Capital and Trade Unions', in I. Mandaza (ed.), *The Political Economy of Transition* CODESRIA (Dakar, 1986)

O. Sichone, 'Pure Anthropology in a Highly Indebted Poor Country', *Journal of Southern African Studies* 27, 2 (2001), pp. 369 - 379

T. Southall, 'Zambia: Class Formation and Government Policy in the 1970s' *Journal of Southern African Studies* 7, 1 (1980), pp. 91 - 108

M. Szeftel, 'Political Graft and the Spoils System in Zambia - the State as a Resource in Itself', *Review of African Political Economy* 24 (1983), pp. 4 - 21

M. Tshoaedi, 'Trade Unions in Zambia', *South Africa Labour Bulletin* 24, 3 (2000), pp. 86 - 90

P. Waterman, 'The 'Labour Aristocracy' in Africa: Introduction to an Unfinished Controversy', in D.L. Cohen & J. Daniel (eds.), *Political Economy of Africa* Longman (London & New York, 1981)

H. Wolpe, 'The Theory of Internal Colonialism – The South African Case' in I. Oxaal, T. Barnett & D. Booth (eds.), *Beyond the Sociology of Development : Economy and Society in Latin America and Africa* Routledge & Kegan Paul (London, 1975)

Unpublished Theses and Other Papers

J. Aron, 'Economic Policy in a Mineral-Dependent Economy: the Case of Zambia', DPhil Thesis, University of Oxford (1992)

D.G. Clarke, 'The Underdevelopment of African Trade Unions in Rhodesia: an Assessment of Working-Class Action in Historical Perspective', PhD Thesis, University of St. Andrews (1974)

J.R. Craig, 'State Enterprise and Privatisation in Zambia, 1968 – 1998' PhD Thesis, University of Leeds (1999)

I. Henderson, 'Labour and Politics in Northern Rhodesia 1900 - 1953: A Study in the Limits of Colonial Power', PhD Thesis, University of Edinburgh (1972)
E. Kalula, 'Labour and the State in Zambia: Towards a View of Post-Colonial Developments', PhD Thesis, University of Warwick (1987)
B.B. Liatto, 'Organised Labour and the State in Zambia', PhD Thesis, University of Leeds (1989)
C.C. Luchembe, 'Finance Capital and Mine Labor: a Comparative Study of Copperminers in Zambia and Peru, 1870-1980', PhD Thesis, University of California (1982)
N.N.J. Mijere, 'The Mineworkers' Resistance to Governmental Decentralisation in Zambia: Nation-Building and Labor Aristocracy in the Third World', PhD Thesis, Brandeis University (1985)
P. Waterman, 'Aristocrats and Plebeians in African Trade Unions?: Lagos Port and Dock Worker Organisation and Struggle', PhD Thesis, Katholieke Universiteit van Nijmegen, Den Haag (1983)
E. Webster & G. Adler, 'Bargained Liberalisation: The Labour Movement Policy Making and the Transition in Zambia and South Africa', Paper presented to the Multinational Working Group on Labor Movement and Policy-making in Africa, Dakar (1997?)
E. West, 'The Politics of Hope: Zambia's Structural Adjustment Programme 1985 – 1987', PhD Thesis, Yale University (1989)

Internet Sources

European Union Election Observation Mission to Zambia 2001, www.eueu-Zambia.org

GLOSSARY

AAC	Anglo American Corporation
ACME	Association of Copper Mining Employers
ALC	Assistant Labour Commissioner
AMWU	African Mineworkers' Union
ANC	African National Congress
BEC	Branch Executive Committee
BGRIMM	Beijing General Research Institute of Mining & Metallurgy
BS	Branch Secretary
CAF	Central African Federation
CCEO	Chairman and Chief Executive Officer
CDC	Commonwealth Development Corporation
CEC	Central Executive Committee (of MUZ)
CIA	Central Intelligence Agency
CID	Criminal Investigation Department
CIRS	Companies' Industrial Relations Secretary
CISB	Copper Industry Service Bureau
CSS	Chief Shop Steward
CSUZ	Civil Servants' Union of Zambia
DG	District Governor
ESAF	Enhanced Structural Adjustment Facility
GC	General Council (of ZCTU)
GDP	Gross Domestic Product
GIRM	Group Industrial Relations Manager
GM	General Manager
GS	General Secretary
ICFTU	International Confederation of Free Trade Unions
INDECO	Industrial Development Corporation
IFI	International Financial Institutions
ILO	International Labour Organisation
IMF	International Monetary Fund
IR	Industrial Relations
IRA	Industrial Relations Act
IRB	Industrial Relations Bureau
JEC	Joint Executive Committee
LC	Labour Commissioner
LME	London Metal Exchange
LO	Labour Officer
LSS	(Ministry of) Labour and Social Security
MAPA	Mines African Police Association
MASA	Mines African Staff Association
MCC	Member of the Central Committee (of UNIP)

MEMACO	Metal Marketing Corporation
MET	Medical Education Trust
MIBPS	Mine Industry Basic Pay Scale
MINDECO	Mining Development Corporation
MJIC	Mines Joint Industrial Council
MLSA	Mines Local Staff Association
MMD	Movement for Multi-Party Democracy
MoL	Ministry of Labour
MPU	Mine Police Unit
MUZ	Mineworkers' Union of Zambia
NAZ	National Archives of Zambia
NC	National Chairman (of MUZ)
NCCM	Nchanga Consolidated Copper Mines
NCR	Ndola Copper Refinery
NERP	New Economic Recovery Programme
NRMWU	Northern Rhodesian Mineworkers' Union
NUBEGW	National Union of Building, Engineering and General Workers' Union
OECD	Organisation for Economic Cooperation and Development
OP	Office of the President
PC	Permanent Committee (of MJIC)
PNT	Privatisation Negotiation Team
Pro MUZ	Professional Mineworkers' Union of Zambia
PRSP	Poverty Reduction Strategy Programme
PS	Permanent Secretary
RAMCoZ	Roan Antelope Mining Company
RCM	Roan Consolidated Mines
RLI	Rhodes-Livingstone Institute
RST	Roan Selection Trust
RTU	Registrar of Trade Unions
RTUC	Reformed Trade Union Congress
SC	Supreme Council (of the MUZ)
SS	Shop Steward
SToZ	Sunday Times of Zambia
TAZARA	Tanzania – Zambia Railway
TLCC	Tripartite Labour Consultative Council
ToZ	Times of Zambia
TUC	Trades Union Congress
UDI	Unilateral Declaration of Independence
UMU	United Mineworkers' Union
UNIP	United National Independence Party
UPP	United Progressive Party
UTUC	United Trade Union Congress
WC	Works Council, Works Councillor
WED	Winding Engine Driver
WFTU	World Federation of Trade Unions
ZANC	Zambia African National Congress
ZCCM	Zambian Consolidated Copper Mines

ZCTU	Zambia Congress of Trade Unions
ZFE	Zambia Federation of Employees
ZIMCO	Zambia Industrial and Mining Corporation
ZMU	Zambia Mineworkers' Union
ZNCB	Zambia National Commercial Bank
ZNCCI	Zambia National Chamber of Commerce and Industry
ZNPF	Zambia National Provident Fund
ZPA	Zambia Privatisation Agency
ZUFIAW	Zambia Union of Financial Institutions and Allied Workers' Union

INDEX

Lightning Source UK Ltd.
Milton Keynes UK
UKHW021832280820
368967UK00006B/83

9 781350 175235